Massively Multilingual Natural Language Understanding (MMNLU-22)

Abu Dhabi, United Arab Emirates
7 December 2022

ISBN: 978-1-7138-9117-8

Printed from e-media with permission by:

Curran Associates, Inc.
57 Morehouse Lane
Red Hook, NY 12571

Some format issues inherent in the e-media version may also appear in this print version.

Printed with permission by Curran Associates, Inc. (2024)

For permission requests, please contact the Association for Computational Linguistics at the address below.

Association for Computational Linguistics
209 N. Eighth Street
Stroudsburg, Pennsylvania 18360

Phone: 1-570-476-8006
Fax: 1-570-476-0860

acl@aclweb.org

Additional copies of this publication are available from:

Curran Associates, Inc.
57 Morehouse Lane
Red Hook, NY 12571 USA
Phone: 845-758-0400
Fax: 845-758-2633
Email: curran@proceedings.com
Web: www.proceedings.com

MMNLU-22 2022

Massively Multilingual Natural Language Understanding 2022

Proceedings of MMNLU-22

December 7, 2022

Introduction

Let's scale natural language understanding technology to every language on Earth!

By 2023 there will be over 8 billion virtual assistants worldwide, the majority of which will be on smartphones. Additionally, over 100 million smart speakers have been sold, most of which exclusively use a voice interface and require Natural Language Understanding (NLU) during every user interaction in order to function. However, even as we approach the point in which there will be more virtual assistants than people in the world, major virtual assistants still only support a small fraction of the world's languages. This limitation is driven by the lack of labeled data, the expense associated with human-based quality assurance, model maintenance and update costs, and more. Innovation is how we will jump these hurdles. The vision of this workshop is to help propel natural language understanding technology into the 50-language, 100-language, and even the 1,000-language regime, both for production systems and for research endeavors.

For an overview of the workshop and competition, please see the paper entitled "The Massively Multilingual Natural Language Understanding 2022 (MMNLU-22) Workshop and Competition," included in these proceedings.

Organizing Committee

Organizers

Jack FitzGerald, Amazon Alexa, USA
Kay Rottmann, Amazon Alexa, Germany
Julia Hirschberg, Columbia University, USA
Mohit Bansal, University of North Carolina, USA
Anna Rumshisky, University of Massachusetts Lowell, USA
Charith Peris, Amazon Alexa, USA
Christopher Hench, Amazon Alexa, USA

Program Committee

Reviewers

Christopher Church

Yulia Grishina

Thanh-Le Ha, He He, Kuan-Hao Huang

Tushar Jain

Philipp Koehn

Thu Le

Yixin Nie, Jan Niehues

Udita Patel

Sayambhu Sen, Pankaj Kumar Sharma, Shubham Shukla, Veselin Stoyanov

Gokhan Tur

Xiang Zhou

Keynote Talk: Fine-grained Multi-lingual Disentangled Autoencoder for Language-Agnostic Representation Learning

Zhongkai Sun
Amazon

Abstract: Encoding both language-specific and language-agnostic information into a single high-dimensional space is a common practice of pre-trained Multi-lingual Language Models (pMLM). Such encoding has been shown to perform effectively on natural language tasks requiring semantics of the whole sentence (e.g., translation). However, its effectiveness appears to be limited on tasks requiring partial information of the utterance (e.g., multi-lingual entity retrieval, template retrieval, and semantic alignment). In this work, a novel Fine-grained Multilingual Disentangled Autoencoder (FMDA) is proposed to disentangle fine-grained semantic information from language-specific information in a multi-lingual setting. FMDA is capable of successfully extracting the disentangled template semantic and residual semantic representations. Experiments conducted on the MASSIVE dataset demonstrate that the disentangled encoding can boost each other during the training, thus consistently outperforming the original pMLM and the strong language disentanglement baseline on monolingual template retrieval and cross-lingual semantic retrieval tasks across multiple languages.

Bio: TBD

Keynote Talk: Towards Efficient Transfer Learning Across Languages

Mahdi Namazifar
Amazon Alexa AI

Abstract: Unbalanced distribution of text resources necessary for AI research and development across languages is well known to result in biases and unfairness in access to benefits of advances in AI. Multilingual language models have played a big role in transferring learnings across languages, facilitating addressing this imbalance to some extend. This talk focuses on additional approaches that could potentially further enable transfer of learnings across languages.

Bio: Mahdi Namazifar received his PhD in Operations Research with a focus in Optimization from University of Wisconsin-Madison in 2011. After PhD he worked at various companies such as Cisco, Twitter, and Uber on applications of machine learning in different industries. In 2020 he joined Amazon Alexa's Conversation Modeling team where he is working on different problems in NLP and Conversational AI.

Keynote Talk: Byte-Level Massively Multilingual Semantic Parsing (Zero-Shot Shared Task Winners)

Massimo Nicosia
Google

Abstract: Token free approaches have been successfully applied to a series of word and span level tasks. In this work, we evaluate a byte-level sequence to sequence model (ByT5) on the 51 languages in the MASSIVE multilingual semantic parsing dataset. We examine multiple experimental settings: (i) zero-shot, (ii) full gold data and (iii) zero-shot with synthetic data. By leveraging a state-of-the-art label projection method for machine translated examples, we are able to reduce the gap in exact match to only 5 points with respect to a model trained on gold data from all the languages. We additionally provide insights on the cross-lingual transfer of ByT5 and show how the model compares with respect to mT5 across all parameter sizes

Bio: Massimo Nicosia is a Senior Software Engineer in Research at Google in Zurich working on making natural language understanding models multilingual.

Keynote Talk: Machine Translation for Multilingual Intent Detection and Slots Filling (Organizers' Choice Award)

Maxime De Bruyn
University of Antwerp

Abstract: We expect to interact with home assistants irrespective of our language. However, scaling the Natural Language Understanding pipeline to multiple languages while keeping the same level of accuracy remains a challenge. In this work, we leverage the inherent multilingual aspect of translation models for the task of multilingual intent classification and slot filling. Our experiments reveal that they work equally well with general-purpose multilingual text-to-text models. Furthermore, their accuracy can be further improved by artificially increasing the size of the training set. Unfortunately, increasing the training set also increases the overlap with the test set, leading to overestimating their true capabilities. As a result, we propose two new evaluation methods capable of accounting for an overlap between the training and test set.

Bio: Maxime is a PhD student in computational linguistics at the University of Antwerp (Belgium) under the supervision of Prof. Walter Daelemans. His work mainly focuses on conversational agents and question answering. Prior to starting his PhD, Maxime was a fund manager at a Belgian private bank.

Keynote Talk: Multilingual NLP for Customer Relationship Management

Géraldine Damnati
Orange Labs

Abstract: Natural Language Processing has become a key technology to improve Customer Relationship Management. Extracting key insights from customer feedbacks, mining opinions from surveys or reviews, designing interactive chatbots or voicebots for commercial or technical assistance are examples among several applications where processing language helps managing Customer Relationship. Being able to handle multiple languages is a central feature for companies, whether when operating in a country where several languages are spoken or when operating in several countries. Recent advances in multilingual NLP represent a huge opportunity towards leveraging customer feedbacks expressed in different languages but many challenges remain.
In this talk, I will present some issues encountered in an international company when analyzing its interactions with customers. In the case of Orange, which also operates in Africa and Middle east, low resource languages are also at stakes. I will address the design of multilingual NLP models in a context where using multipurpose Large Language Models or even any model needing GPU computation is not always a realistic scalable solution. I will share experience of data collection in the context of highly regulated domain with European General Data Protection Regulation and of data annotation in a context where micro-tasking is generally not used. I will also discuss how to bridge the gap between academic research on unconstrained benchmark corpora that do not always fit the reality of deployment constraints and how these constraints can fuel new research questions.

Bio: Géraldine Damnati is a Research Engineer at Orange Innovation, DATA&AI, Lannion, France. After an engineering degree from Telecom Bretagne, she obtained in 2000 a PhD in Computer Science from University of Avignon. Her research interests include Natural Language Processing, Spoken Language Understanding, Text and Speech Mining, Semantic Analysis, Question Answering and Information Extraction in general. She has a research activity, contributing to collaborative projects, being co-author of around 80 publications in international conferences. She is also involved in the conception and development of tools in various applicative domains, such as Customer Relationship or Multimedia Content exploration. She is currently involved in several research projects, including the ARCHIVAL pluridisciplinary project (http://archival.msh-paris.fr/) for archive valorisation in the context of Digital Humanities. She is also active in the French NLP community, as a member of the ATALA board (Association pour le Traitement Automatique des Langues) and as the coordinator of the French CNRS GDR-TAL partners club.

Keynote Talk: Towards Massively Multilingual Modular Models

Sebastian Ruder
Google

Abstract: State-of-the-art multilingual models are trained on data of around 100 languages. These models can be adapted to perform better in under-represented languages but such adaptation does not directly benefit the original models. In order to make progress on NLU capabilities for the next 1,000 languages, we need to make it easier for researchers from diverse backgrounds to build upon and share improvements on base models. To this end, I will first discuss the tools currently at our disposal for extending multilingual models, from sparse subnetworks to parameter-efficient adaptation and vocabulary extension. I will then highlight the benefits of modularity compared to current model monoliths. Finally, I will sketch a vision of how we can build, train, and evaluate modular multilingual models that can cover the next 1,000 languages.

Bio: Sebastian is a research scientist at Google based in Berlin, Germany working on natural language processing (NLP) for under-represented languages. Before that he was a research scientist at DeepMind. He completed his PhD in Natural Language Processing and Deep Learning at the Insight Research Centre for Data Analytics, while working as a research scientist at Dublin-based text analytics startup AYLIEN. Previously, he studied Computational Linguistics at the University of Heidelberg, Germany and at Trinity College, Dublin. He's interested in cross-lingual and transfer learning for NLP and making ML and NLP more accessible.

Keynote Talk: HIT-SCIR at MMNLU-22: Consistency Regularization for Multilingual Spoken Language Understanding (Best Paper and Full-Data Shared Task Winner)

Bo Zheng
Research Center for Social Computing and Information Retrieval (SCIR) of Harbin Institute of Technology

Abstract: Multilingual spoken language understanding (SLU) consists of two sub-tasks, namely intent detection and slot filling. To improve the performance of these two sub-tasks, we propose to use consistency regularization based on a hybrid data augmentation strategy. The consistency regularization enforces the predicted distributions for an example and its semantically equivalent augmentation to be consistent. We conduct experiments on the MASSIVE dataset under both full-dataset and zero-shot settings. Experimental results demonstrate that our proposed method improves the performance on both intent detection and slot filling tasks. Our system ranked 1st in the MMNLU-22 competition under the full-dataset setting.

Bio: Bo Zheng is a final-year Ph.D. student at the Research Center for Social Computing and Information Retrieval (SCIR) of Harbin Institute of Technology, advised by Prof. Wanxiang Che. His research interests include cross-lingual NLP, machine reading comprehension, and language analysis. He has published many papers in international conferences and journals such as ACL, EMNLP, CoNLL, etc. He was ranked first on multiple official leaderboards, including the leaderboard of Google's XTREME benchmark, Google's Natural Questions dataset, and Amazon's MASSIVE dataset. He was also ranked first in multiple international competitions, including Amazon's MMNLU-2022 competition, CoNLL 2018 shared task, and NLP-TEA 2016 shared task.

Keynote Talk: Massively Multilingual NLP in 1600+ Languages

David Yarowsky
Johns Hopkins University

Abstract: The talk will cover a range of topics in massively multilingual and very low-resource NLP and speech recognition, in core functionalities, at a nearly unprecedented language-universal scale.

Bio: David Yarowsky is a Professor of Computer Science at Johns Hopkins University, and a member of its Center for Language and Speech Processing. He received his PhD from the University of Pennsylvania in 1996. He is an ACL Fellow, NSF CAREER award winner, Rockefeller Fellow, summa-cum-laude graduate from Harvard, ACL Test-of-time award winner, ACL Treasurer, co-founder of the EMNLP conference series and longtime ACL/SIGDAT executive committee member. He has pioneered the field of cross-lingual information projection via bilingual word alignments and done extensive work in low-resource and massively multilingual NLP, and is also known for an eponymous influential algorithm used for co-training, multi-view machine learning and low-resource bootstrapping.

Keynote Talk: Learning in the Wild: Modeling Language in Real-World Scenarios

Anna Rumshisky
University of Massachusetts Lowell

Abstract: Scientific progress in NLP is often measured by model performance on standardized benchmarks. But in many cases, existing benchmarks fail to reflect the settings in which algorithmic solutions are applied in practice. The challenges of modeling language in real-world scenarios often go beyond covariate shift and related well-studied phenomena. In this talk, I will discuss some of these challenges, using user interactions with digital assistants as a case study. I will describe some recent work aimed at addressing such challenges, including (a) learning from a combination of positive and negative noisy user feedback in a federated setting, and (b) learning from frequency-enriched data in a setting where a different treatment is required for the head and tail of the distribution.

Bio: Anna Rumshisky is an Associate Professor of Computer Science at the University of Massachusetts Lowell, where she heads the Text Machine Lab for NLP. Her primary research area is machine learning for natural language processing, with a focus on deep learning techniques. She has made contributions in a number of application areas, including computational lexical semantics, temporal reasoning and argument mining, as well as clinical informatics and computational social science. She received her PhD from Brandeis University and completed postdoctoral training at MIT CSAIL, where she is currently a Research Affiliate. She is a recipient of the NSF CAREER award in 2017, and her work won the best thematic paper award at NAACL-HLT 2019.

Keynote Talk: Multilingual Information Extraction for Thousands of Types

Heng Ji
University of Illinois at Urbana-Champaign

Abstract: Supervised information extraction models require a substantial amount of training data to perform well. However, information annotation requires a lot of human effort and costs much time, especially for low-resource languages, which limits the application of existing supervised approaches to new knowledge types. In order to reduce manual labor and shorten the time to build an information extraction system for an arbitrary ontology, we present a new framework to train such systems much more efficiently without large annotations. Our weak supervision approach only requires a set of keywords, a small number of examples and an unlabeled corpus in any language, and takes advantage of naturally existing "hubs" (such as linking to WikiData, Multilingual embedding and universal semantic parsers) for cross-lingual transfer.

Bio: Heng Ji is a professor at Computer Science Department, and an affiliated faculty member at Electrical and Computer Engineering Department of University of Illinois at Urbana-Champaign. She is an Amazon Scholar. She received her B.A. and M. A. in Computational Linguistics from Tsinghua University, and her M.S. and Ph.D. in Computer Science from New York University. Her research interests focus on Natural Language Processing, especially on Multimedia Multilingual Information Extraction, Knowledge Base Population and Knowledge-driven Generation. She was selected as Young Scientist and a member of the Global Future Council on the Future of Computing by the World Economic Forum in 2016 and 2017. The awards she received include AI's 10 to Watch Award by IEEE Intelligent Systems in 2013, NSF CAREER award in 2009, Google Research Award in 2009 and 2014, IBM Watson Faculty Award in 2012 and 2014, Bosch Research Award in 2014-2018, Best-of-ICDM2013 Paper, Best-of-SDM2013 Paper, ACL2020 Best Demo Paper Award, and NAACL2021 Best Demo Paper Award. She is elected as the North American Chapter of the Association for Computational Linguistics (NAACL) secretary 2020-2021. She has served as the Program Committee Co-Chair of many conferences including NAACL-HLT2018, and she has been the coordinator for the NIST TAC Knowledge Base Population track since 2010.

Table of Contents

Program

Wednesday, December 7, 2022

09:00 - 09:30	*Introduction and Shared Task Overview*
09:30 - 10:00	*Fine-grained Multi-lingual Disentangled Autoencoder for Language-Agnostic Representation Learning*
10:00 - 10:30	*Invited Talk by Mahdi Namazifar: Towards Efficient Transfer Learning Across Languages*
10:30 - 11:00	*Break*
11:00 - 11:30	*Zero-Shot Shared Task Winners: Massimo Nicosia and Francesco Piccinno, Google*
11:30 - 12:00	*Invited Talk by Sebastian Ruder, Google: Towards Massively Multilingual Modular Models*
12:00 - 12:30	*Invited Talk by Géraldine Damnati, Orange Labs: Multilingual NLP for Customer Relationship Management*
12:30 - 13:30	*Lunch*
13:30 - 14:00	*Organizers' Choice Award: Maxime De Bruyn and the bolleke team*
14:00 - 14:30	*Best Paper Award and Full-Data Shared Task Winner: Bo Zheng and the HIT-SCIR team*
14:30 - 15:30	*Poster Session*
15:30 - 16:00	*Break*
16:00 - 16:30	*Invited Talk by David Yarowsky, JHU: Massively Multilingual NLP in 1600+ Languages*
16:30 - 17:00	*Invited Talk by Anna Rumshisky, UMass Lowell: Learning in the Wild: Modeling Language in Real-World Scenarios*
17:00 - 17:30	*Invited Talk by Heng Ji, U of Illinois Urbana-Champaign: Multilingual Information Extraction for Thousands of Types*
17:30 - 18:30	*Networking*

Robust Domain Adaptation for Pre-trained Multilingual Neural Machine Translation Models

Mathieu Grosso, Pirashanth Ratnamogan, Alexis Mathey,
William Vanhuffel, Michael Fotso Fotso
BNP Paribas
(mathieu.grosso, pirashanth.ratnamogan, alexis.mathey, william.vanhuffel,michael.fotsofotso)@bnpparibas.com

Abstract

Recent literature has demonstrated the potential of multilingual Neural Machine Translation (mNMT) models. However, the most efficient models are not well suited to specialized industries. In these cases, internal data is scarce and expensive to find in all language pairs. Therefore, fine-tuning a mNMT model on a specialized domain is hard. In this context, we decided to focus on a new task: *Domain Adaptation of a pre-trained mNMT model on a single pair of language* while trying to maintain model quality on generic domain data for all language pairs. The risk of loss on generic domain and on other pairs is high. This task is key for mNMT model adoption in the industry and is at the border of many others. We propose a fine-tuning procedure for the generic mNMT that combines embeddings freezing and adversarial loss. Our experiments demonstrated that the procedure improves performances on specialized data with a minimal loss in initial performances on generic domain for all languages pairs, compared to a naive standard approach (+10.0 BLEU score on specialized data, -0.01 to -0.5 BLEU on WMT and Tatoeba datasets on the other pairs with M2M100).

1 Introduction

Building a NMT model supporting multiple language pairs is an active and emerging area of research (NLLB Team et al., 2022; Fan et al., 2020; Tang et al., 2020). Multilingual NMT(mNMT) uses a single model that supports translation in multiple language pairs. Multilingual models have several advantages over their bilingual counterparts (Arivazhagan et al., 2019b). This modeling proves to be both efficient and effective as it reduces the operational cost (a single model is deployed for all language pairs) and improves translation performances, especially for low-resource languages.

All these advantages make mNMT models interesting for real-world applications. However, they are not suitable for specialized industries that require domain-specific translation. Training a model from scratch or fine-tuning all the pairs of a pre-trained mNMT model is almost impossible for most companies as it requires access to a large number of resources and specialized data. That said, fine-tuning a single pair of a pre-trained mNMT model in a specialized domain seems possible. Ideally this domain adaptation could be learned while sharing parameters from old ones, without suffering from catastrophic forgetting (Mccloskey and Cohen, 1989). This is rarely the case. The risk of degrading performance on old pairs is high due to the limited available data from the target domain and to the extremely high complexity of the pre-trained model. **In our case, overfitting on fine-tuning data means that the model might not even be multilingual anymore**

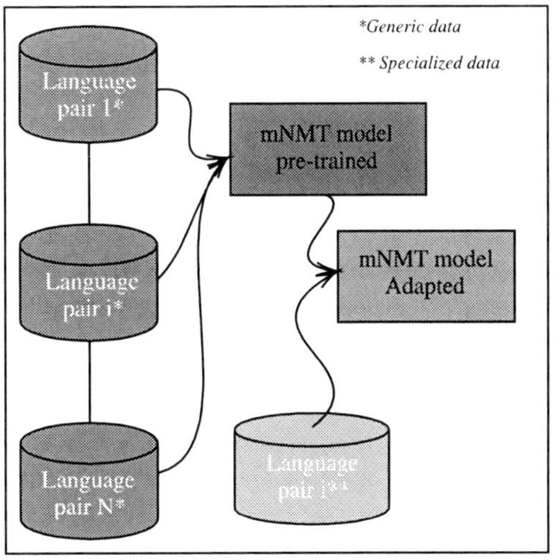

Figure 1: Domain Adaptation of a Pre-trained mNMT

In this context, this article focuses on a new real-world oriented task **fine-tuning a pre-trained mNMT model in a single pair of language on a specific domain without losing initial performances on the other pairs and generic data**. Our

Proceedings of the Massively Multilingual Natural Language Understanding Workshop (MMNLU-22), pages 1 - 11
December 7, 2022 ©2022 Association for Computational Linguistics

research focuses on fine-tuning two state-of-the-art pre-trained multilingual mNMT freely available: M2M100 (Fan et al., 2020) and mBART50 (Tang et al., 2020) which both provide high performing BLEU scores and translate up to 100 languages.

We explored multiple approaches for this domain adaptation. Our experiments were made on English to French data from medical domain[1]. This paper shows that fine-tuning a pre-trained model with initial layers freezing, for a few steps and with a small learning rate is the best performing approach.

It is organized as follows : firstly, we introduce standard components of modern NMT, secondly we describe related works, thirdly we present our methods. We finally systematically study the impact of some state-of-the-art fine-tuning methods and present our results.

Our main contributions can be separated into 2 parts:

- Defining a new real-world oriented task that focuses on domain adaptation and catastrophic forgetting on multilingual NMT models
- Defining a procedure that allows to finetune a pre-trained generic model on a specific domain

2 Background

2.1 Neural Machine Translation

Neural Machine Translation (NMT) has become the dominant field of machine translation. It studies how to automatically translate from one language to another using neural networks.

Most NMT systems are trained using Seq2Seq architectures (Sutskever et al., 2014; Cho et al., 2014) by maximizing the prediction of the target sequence $V_T = (v_1, \ldots, v_T)$, given the source sentence $W_S = (w_1, \ldots, w_S)$:

$$P\left(v_1, \ldots, v_T \mid w_1, \ldots, w_S\right)$$

Today the best performing Seq2Seq architecture for NMT is based on Transformers (Vaswani et al., 2017) architecture. They are built on different layers among which the multi-head attention and the feed-forward layer. These are applied sequentially and are both followed by a residual connection (He et al., 2015) and layer normalization (Ba et al., 2016).

Although powerful, traditional NMT only translates from one language to another with a high computational cost compared to its statistical predecessor. It has been shown that a simple language token can condition the network to translate a sentence in any target language from any source language (Johnson et al., 2017). It allows to create multilingual models that can translate between multiple languages. Using previous notation the multilingual model adds the condition on target language in the previous modeling

$$P\left(v_1, \ldots, v_T \mid w_1, \ldots, w_S, \ell\right)$$

where ℓ is the target language.

2.2 Transfer Learning

Transfer learning is a key topic in Natural Language Processing (Devlin et al., 2018; Liu et al., 2019). It is based on the assumption that pre-training a model on a large set of data in various tasks will help initialize a network trained on another task where data is scarce.

It is already a key area of research in NMT where large set of generic data are freely available (news, common crawl, ...). However, real-world applications require specialized models. In-domain data is rare and more costly to gather for industries (finance, legal, medical, ...) making specialized models harder to train. It is even more true for multilingual model.

In our work, we study how we can adapt a mNMT model on a specific domain by fine-tuning on only one language pair, without losing too much generality for all language pairs.

3 Related works

3.1 Multilingual Neural Machine Translation

While initial research on NMT started with bilingual translation systems (Sutskever et al., 2014; Cho et al., 2014; Luong et al., 2015; Yang et al., 2020), it has been shown that the NMT framework is extendable to multilingual models (Dong et al., 2015; Firat et al., 2016; Johnson et al., 2017; Dabre et al., 2020) mNMT has seen a sharp increase in the number of publications, since it is easily extendable and it allows both end-to-end modeling and cross

[1] https://opus.nlpl.eu/EMEA-v3.php

lingual language representation (Conneau et al., 2017; Linger and Hajaiej, 2020; Conneau et al., 2019).

Competitive multilingual models have been released and open sourced. mBART (Liu et al., 2019) first, was trained following the BART (Lewis et al., 2019) objective before being finetuned on an English-centric multilingual dataset (Tang et al., 2020). M2M100 (Fan et al., 2020) scaled large transformer layers (Vaswani et al., 2017) with a lot of mined data in order to create a mNMT without using English as pivot, that can perform translation between any pairs among 100 languages. More recently, NLLB was released (NLLB Team et al., 2022), extending the M2M100 framework to 200 languages. Those models are extremely competitive as they have similar performance to their bilingual counterpart while allowing a pooling of training and resources.

Our experiments will rely on M2M100 and mBART but it can be generalized to any new pre-trained multilingual model (NLLB Team et al., 2022).

3.2 Domain Adaptation

Domain Adaptation in the field of NMT is a key real-world oriented task. It aims at maximizing model performances on a certain in-domain data distribution. Dominant approaches are based on fine-tuning a generic model using either in-domain data only or a mixture of out-of-domain and in-domain data to reduce overfitting (Servan et al., 2016a; Van Der Wees et al., 2017). Many works have extended domain adaptation to multi-domain, where model is finetuned on multiple and different domains (Sajjad et al., 2017; Zeng et al., 2018; Mghabbar and Ratnamogan, 2020).
However, to the best of our knowledge, our work is the first exploring domain adaptation in the context of recent pre-trained multilingual neural machine translation systems, while focusing on keeping the model performant in out-of-domain data in all languages.

3.3 Learning without forgetting

Training on a new task or new data without losing past performances is a generic machine learning task, named Learning without forgetting (Li and Hoiem, 2016).

Limiting pre-trained weights updates using either trust regions or adversarial loss is a recent idea that has been used to improve training stability in both natural language processing and computer vision (Zhu et al., 2019; Jiang et al., 2020; Aghajanyan et al., 2020). These methods haven't been explored in the context of NMT but are key assets that demonstrated their capabilities on other NLP tasks (Natural Language Inference in particular). Our work will apply a combination of those methods to our task.

3.4 Zero Shot Translation

MNMT has shown the capability of direct translation between language pairs unseen in training: a mNMT system can automatically translate between unseen pairs without any direct supervision, as long as both source and target languages were included in the training data (Johnson et al., 2017). However, prior works (Johnson et al., 2017; Firat et al., 2016; Arivazhagan et al., 2019a) showed that the quality of zero-shot NMT significantly lags behind pivot-based translation (Gu et al., 2019). Based on these ideas, some paper (Liu et al., 2021) have focused on training a mNMT model supporting the addition of new languages by relaxing the correspondence between input tokens and encoder representations, therefore improving its zero-shot capacity. We were interested in using this method as learning less specific input tokens during the finetuning procedure could help our model not to overfit the training pairs. Indeed, generalizing to a new domain can be seen as a task that includes generalizing to an unseen language.

4 Methods

Our new real-world oriented task being at the crossboard of many existing task, we applied ideas from current literature and tried to combine different approaches to achieve the best results.

4.1 Hyperparameters search heuristics for efficient fine-tuning

We seek to adapt generic multilingual model to a specific task or domain. (Cettolo et al., 2014; Servan et al., 2016b). Recent works in NMT (Domingo et al., 2019) have proposed methods to adapt incrementally a model to a specific domain. We continue the training of the generic model on specific data, through several iterations (see Algorithm 1). This post-training fine-tuning procedure is done without dropping the previous learning states of the multilingual model. The resulting model is considered as adapted or specialized to a specific domain. We want to avoid the model to suffer from forgetting on generic domain and pairs. To this end, we include different methods in this fine-tuning, that have been mentioned in the literature. These methods includes in particular choosing a small learning rate (Howard and Ruder, 2018), a triangular learning schedule (Houlsby et al., 2019), reducing the number of steps and freezing some of the layers(Stickland and Murray, 2019).

4.2 Smoothness-inducing Adversarial Regularizer

We seek to reduce the loss on generic domain and other pairs. Indeed, due to limited data resources from downstream tasks and the extremely large capacity of pre-trained models, aggressive fine-tuning often causes the adapted model to overfit the data of downstream tasks and forget the knowledge of the pre-trained model. To this end, we added a Smoothness-inducing Adversarial Regularization (SMART) term during the fine-tuning (Jiang et al., 2020). Models fine-tuned on GLUE task with SMART approach outperform even the strongest pre-trained baseline on all 8 tasks. Comparing with BERT (Devlin et al., 2018) and RoBERTa (Liu et al., 2019), BERT_{SMART} and RoBERTa_{SMART} are performing better by a big margin. This approach gives a smoothness-inducing property to the model f. This is helpful to prevent overfitting and to improve generalization on low resource target domain for a certain task. Therefore, adding it to our task should avoid overfitting on the new domain.

Given the model $f(.;\theta)$ and n data points of the target task denoted by $\{(x_i, y_i)\}_{i=1}^{n}$, where x_i's denote the embeddings of the input sentences, given by the first embedding layer of the language model and y_i's are the associated labels, SMART is adding a regularization term $\mathcal{R}_s(\theta)$ to the canonical optimisation loss below:

$$\min_{\theta}(\mathcal{F}(\theta)) = \mathcal{L}(\theta) + \lambda_s \mathcal{R}_s(\theta) \quad (1)$$

where $\mathcal{L}(\theta)$ is the loss function defined as

$$\mathcal{L}(\theta) = \frac{1}{n}\sum_{i=1}^{n} \ell\left(f\left(x_i;\theta\right), y_i\right) \quad (2)$$

and $\ell(\cdot,\cdot)$ is the loss function depending on the target task, $\lambda_s > 0$ is a tuning parameter, and $\mathcal{R}_s(\theta)$ is the smoothness-inducing adversarial regularizer.

Here we define $\mathcal{R}_s(\theta)$ as

$$\mathcal{R}_s(\theta) = \frac{1}{n}\sum_{i=1}^{n} \max_{\|\bar{x}_i - x_i\|_p \leq \epsilon} \ell_s\left(f\left(\bar{x}_i;\theta\right), f\left(x_i;\theta\right)\right) \quad (3)$$

where $\epsilon > 0$ is a tuning parameter. Since NMT is a classification tasks, $f(;\theta)$ outputs a probability simplex and ℓ_s is chosen as the symmetrized KL-divergence, i.e.,

$$\ell_s(P, Q) = \mathcal{D}_{\mathrm{KL}}(P\|Q) + \mathcal{D}_{\mathrm{KL}}(Q\|P)$$

4.3 Enabling the model to learn less aggressive input tokens

We seek at reducing the loss of performances on the pairs learned during the pre-training of the model. A factor causing a too important language-specific representation is the positional correspondence to input tokens (Liu et al., 2021). Relaxing it should help the model learn the new domain while not focusing too much on the language representation. Recent advances in mNMT showed that we can reduce the positional correspondence learned from the input tokens seen during training thanks to Positional Disentangling Encoder (PDE) (Liu et al., 2021). PDE corresponds to removing some of the residual connections of the model architecture. PDE is reported to beat by +18.5 BLEU models that do not use it on zero shot translation pairs while retaining quality on supervised directions (Liu et al., 2021). Doing this during the domain adaptation fine-tuning helps to learn less specific input tokens (since we train only from English to French). Therefore, adapting this method to our domain adaptation training is straightforward and could bring gain in BLEU on language pairs seen during pre-training while not sacrificing performances on the new specific domain.

5 Experimental Settings

5.1 Pre-trained Generic Models used

We have worked with two pre-trained mNMT models: M2M100 and mBART50 large.

M2M100 is a multilingual encoder-decoder model, based on large Transformer architecture that can handle 100 languages. It was trained on a non-English-centric dataset of 7.5B sentences from generic domain, as such it is the first true many-to-many NMT model. To ease the fine-tuning process and due to hardware limitations, we worked with the lightest version released (418M parameters).

mBART50 is a multilingual encoder-decoder model, based on training on an English-centric dataset and on large Transformer architecture that can handle 50 languages. It was trained following the BART objective (Lewis et al., 2019). More formally, the model aims to reconstruct a text that has been previously noised.

We will compare the domain adaptation performance between mBART50 which was trained on English-centric data and M2M100 which was trained on non English-centric data.

5.2 Datasets and preprocessing

In order to assess the effectiveness of our different domain adaptation strategies, we focused on the medical domain on the **English to French** using data from the EMEA3[2] dataset (Tiedemann, 2012). We used the same preprocessing as the original publications (BPE joint-tokenization from sentencepiece). We split the dataset into a train and a test dataset. We chose to use the first 5.000 sentences for the testing set and 350.000 sentences for the training set. For the evaluation data on the generic domain, we used generic data from different sources including WMT[3] and Tatoeba[4]. For the evaluation data on the medical domain, we also used EMEA3 dataset in different languages.

5.3 Detailed Procedure

We first define a hyperparameters search heuristics procedure. We chose a range of learning rate and trained the model with these values. We set prior threshold between the loss we accept on generic data and the increase we target on medical data. Then apply the procedure in algorithm 1. Having done this, we kept best settings (best learning rate and number of steps for given threshold), and tried freezing first layers to reduce the loss on generic domain. We define ϵ_3, a threshold between loss on medical domain and gain on generic domain. We reproduce the same procedure and reports our best results. This allows us to find the optimal model θ_{opt}, representing the best compromise between not losing performances on generic data and good adaptation to the medical domain.

[2]https://opus.nlpl.eu/EMEA.php

[3]https://opus.nlpl.eu/WMT-News.php

[4]https://github.com/Helsinki-NLP/Tatoeba-Challenge

Algorithm 1 Hyperparameters search heuristic for domain adaptation using simple fine-tuning Algorithm

Input: T : the maximum number of steps; L : the number of layers we have frozen; L_r: the learning rate, ϵ_1: the threshold for Δ_1 : the difference of BLEU between baseline and adapted model on EN-FR generic domain data, ϵ_2 threshold for Δ_2 : the mean difference of BLEU between baseline and adapted model on all other generic data, θ_0 is the parameters of the pretrained model, θ_{opt}: is the parameters of the model that has optimal value of BLEU on domain and generic.

$T \leftarrow 100K$
$L \leftarrow 1$
for $L_r = 3e-5, 1e-5, ..., 1e-8$ **do**
  $\theta_s \leftarrow \theta_0$
  for $s \leftarrow 1$ to T **do**
    $\theta_{s+1} \leftarrow \text{AdamUpdate}_B(\theta_s)$
    Every 2k steps, evaluate model on validation set and compute Δ_1 and Δ_2
    if $\Delta_1 \leq \epsilon_1 \cup \Delta_2 \leq \epsilon_2$ is true **then**
      $\theta_{opt} \leftarrow \theta_s$
    else
      $\theta_{opt} \leftarrow \theta_s$
      end For loop
    end if
  end for
end for

Output: θ_{opt}

M2M100 We trained M2M100 on the medical EN-FR dataset. We used the adam optimizer ($\beta_1 = 0.9, \beta_2 = 0.98$), label smoothing, a dropout of 0.1 and a weight decay of 0. We applied our hyperparameters search heuristic procedure 1 to find the best model. We set $\epsilon_1 = 2, \epsilon_2 = 1$. On this configuration, optimal results were reported with a learning rate of 1e-07, freezing the embeddings at the encoder level, and 60K steps.

mBART50 We trained mBART50 large on the medical EN-FR dataset. We used the adam optimizer ($\beta_1 = 0.9, \beta_2 = 0.98$), label smoothing, a dropout of 0.3 and a weight decay of 0. Again, we applied our hyperparameters search heuristic procedure to find the best model 1. We had to increase the value of ϵ_1, ϵ_2 since mBART50 tends to forget the generic domain quicker than M2M100. We set $\epsilon_1 = 4, \epsilon_2 = 3$. On this configuration, optimal results were reported with a learning rate of $6e-07$, freezing the embeddings at the encoder level and 10K steps.

SMART: We finetuned the model with the SMART procedure and continue hyperparameters search as in algorithm 1. In Algorithm 2, we note $\mathcal{R}_s(\theta) = \frac{1}{|B|}\sum_{x_i \in B} \max_{\|\bar{x}_i - x_i\|p \leq \epsilon} \ell_s\left(f\left(\bar{x}_i;\theta\right), f\left(x_i;\theta\right)\right)$ and $AdamUpdate$ the ADAM update rule for optimizing equation 1 using the mini-batch B. Lastly, we set $T_{\bar{x}} = 1$. For the perturbation, we set $\epsilon = 10^{-5}$ and $\sigma = 10^{-5}$. The learning rate η is set to 10^{-3}.

Algorithm 2 Adding SMART to procedure

Input: T : the total number of iterations; $\mathcal{X}$: the dataset; θ_0: the parameter of the pre-trained model; σ^2: the variance of the random initialization for $\bar{x}_i$'s; $T_{\bar{x}}$: the number of iterations for updating $\bar{x}_i$'s; η: the learning rate for updating $\bar{x}_i$'s; β: clipping value.

$\theta_1 \leftarrow \theta_0$
for $t \leftarrow 1$ to T **do**
  $\bar{\theta}_s \leftarrow \theta_t$
  Sample a mini-batch B from $\mathcal{X}$
  For all $x_i \in B$, initialize $\bar{x}_i \leftarrow x_i + v_i$ with $v_i \sim \mathcal{N}\left(0, \sigma^2 I\right)$
  for $m = 1, ..., T_{\bar{x}}$ **do**
    $\bar{x}_i \leftarrow \bar{x}_i + \eta \mathcal{R}_s(\bar{\theta}_s)$
  end for
  $\bar{\theta}_{s+1} \leftarrow AdamUpdate_B\left(\bar{\theta}_s\right)$
  $\theta_{t+1} \leftarrow CLIP(\bar{\theta}_{s+1}, 1-\beta, 1+\beta)$
end for

Output: θ_T

PDE Finally, we define PDE. It consists in applying Algorithm 1 and then removing first all the residual connection in the penultimate Encoder layers (Chen et al., 2022), then we try removing only the attention layer residual connections (figure 2).

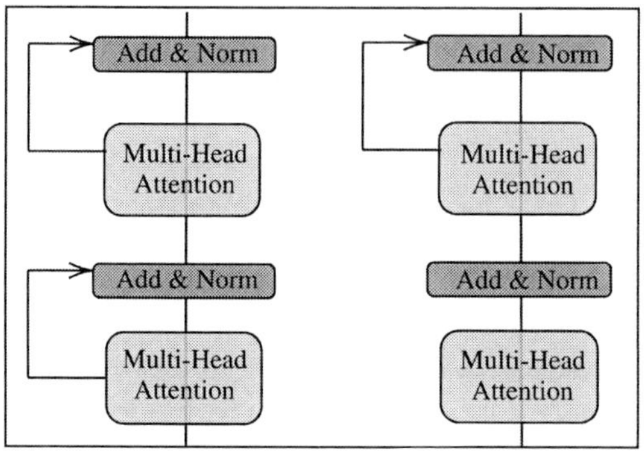

Figure 2: PDE Illustration: Removing Residual Connections on encoder block

6 Results and Analysis

6.1 Hyperparameters search heuristic

6.1.1 Main Results

M2M100 As shown in table 1 we reached more than 9.00 increase of BLEU score on the medical dataset without sacrificing performance on generic domain, the loss is not important on most of the pairs (between 0.01 and 0.2). In figure 3, we see that the mean results is rather stable and that the BLEU on generic English to French data does not decrease a lot (around -1.5 BLEU). The model converges after 60K steps so we stop training.

mBART50 Again we reach more than 9.00 BLEU increase (Figure 4). We observe that after 50K steps mBART50 starts converging around 40.00 BLEU, yet we decided to stop domain adaptation training sooner than with M2M100 as a trade-off between good performance on the EN-FR medical domain and loss of performance on the generic domain. Globally, we achieved better results with M2M100 than mBART50.

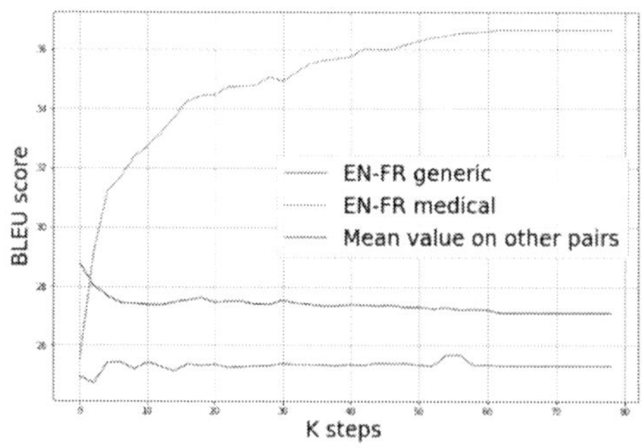

Figure 3: Domain Adaptation (Medical Domain EN-FR) of M2M100

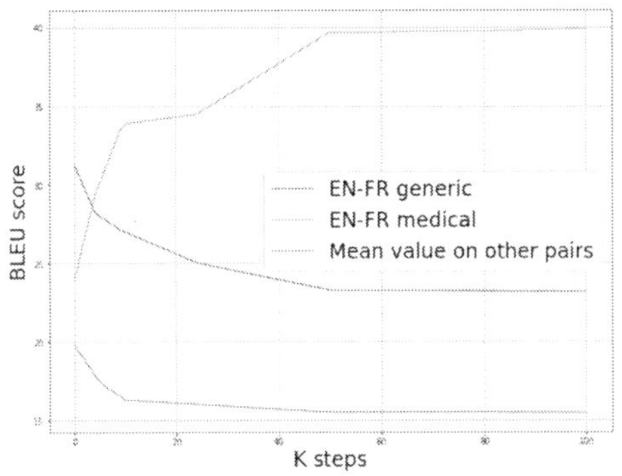

Figure 4: Domain Adaptation (Medical Domain EN-FR) of mBART50

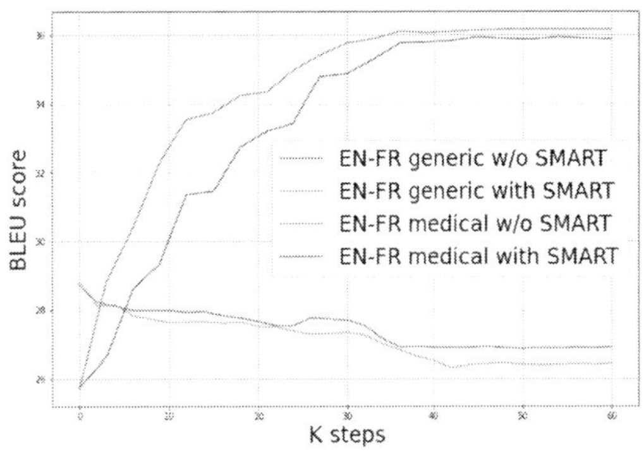

Figure 6: Adding SMART to M2M100 Domain Adaptation training

6.1.2 Catastrophic forgetting with a big learning rate

We tested several learning rate values and we report here our results with a bigger learning rate ($3e-5$). For both models, it led to a catastrophic forgetting on the non-finetuned pairs along with a huge performance increase on the EN-FR Medical dataset, reaching a higher BLEU on the Medical dataset. We decided to focus on a smaller learning rate as a trade-off between loss on generic domain and gain on the medical domain.

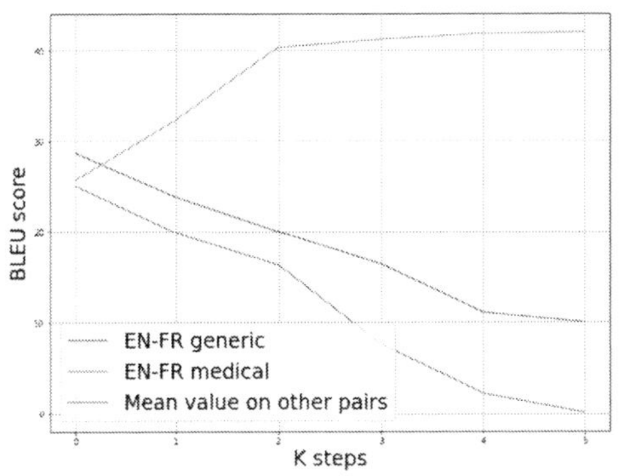

Figure 5: Domain Adaptation of M2M100 with big Learning Rate

6.2 SMART

We have reported our fine-tuning results for M2M100 and mBART50 with SMART in Table 1.

Our goal with SMART was to reach a higher BLEU score on the generic domain data without sacrificing performances on the medical dataset. In Table 1, we note a good increase in BLEU score. Moreover, we have noted that the BLEU change less when moving learning rate in a reasonable range compared to the other methods that are extremely sensitive to hyperparameters. In this context, SMART is useful in order to **achieve quick adaptation of a mNMT model to a new domain. It makes domain adaptation procedure more consistent**. Therefore, SMART training procedure allows efficient and robust domain adaptation. However if exploring a large scale of hyper parameters if feasible simple fine-tuning procedure like in Algorithm 1 can provide better results as shown in Table 1.

6.3 PDE

We seek at reducing the loss of performances on the pairs learned during pre-training of the model (and that are not used during the post-training domain adaptation). Relaxing the correspondence to the input tokens learned during Domain Adaptation. Fine-tuning was supposed to help learning less specific input tokens and therefore the model would be less likely to forgot all the pretrained pairs. As expected, the model learned less aggressive input tokens and do not overfit on English input tokens. However, in practice this does not seem to work well. Indeed, the model is also likelier to forget the pretrained input tokens making this method unfit to our procedure. Using PDE a posteriori (during fine-tuning) seems to be inefficient, since the model is performing worse on all pairs and not only on the English pairs.

We report our results in table 1.

6.4 Analysis

6.4.1 Zero-shot Domain Adaptation on other pairs

We challenged the approach on domain adaptation on languages unseen during the post-training on the

Table 1: Global results on domain adaptation of M2M100 and mBART50

	M2M100				mBART50			
	Baseline	Finetuned	Finetuned SMART	Finetuned PDE	Baseline	Finetuned	Finetuned SMART	Finetuned PDE
EN-FR medical data	26.94	**36.05**	35.93	29.71	23.99	**33.87**	30.3	26.12
EN-FR (WMT)	**28.63**	26.90	26.41	25.01	**31.25**	27.10	26.10	17.56
Mean results	24.97	**25.38**	25.15	21.70	**19.83**	16.85	15.1	13.63
DE-EN (WMT)	22.42	**22.53**	22.39	21.15	**26.13**	21.85	20.64	19.44
EN-DE (WMT)	19.48	**19.52**	19.21	17.96	**22.72**	19.84	18.61	16.83
RU-EN (WMT)	**26.3**	26.27	25.4	24.90	**29.72**	24.64	23.55	22.47
FR-DE (WMT)	**17.82**	17.80	17.58	14.69	**10.92**	8.98	7.1	4.30
EN-FI (WMT)	12.51	**12.72**	12.51	11.74	**13.39**	11.23	10.27	9.74
FI-EN (WMT)	**23.55**	23.18	23.39	21.40	**22.10**	18.90	17.75	16.33
BG-IT (Tatoeba)	26.65	**27.54**	27.01	26.20	*	*	*	*
DA-TR (Tatoeba)	20.22	**22.27**	21.75	20.23	*	*	*	*
PL-RU (Tatoeba)	**33.79**	33.72	33.68	29.69	**14.45**	10.49	10.34	9.87
PT-ES (Tatoeba)	51.49	51.98	**52.54**	50.34	**21.57**	18.87	16.80	12.54
JA-ES (Tatoeba)	20.55	**21.66**	21.58	20.30	**17.55**	15.42	13.27	11.12

medical domain using EMEA3 dataset available on other languages. Table 2 shows that for M2M100 all BLEU scores are increasing, moreover the pairs that implies either English or French are particularly benefiting from this domain adaptation. On mBART50, we also note improvements, first the loss is less important than on generic dataset for the pairs that do not include French as output showing that the model is learning a bit. When French is the output, the domain adaptation is working really fine and we see improvements. Domain-specific data are often hard to gather, especially for low-ressource pairs. That's why being able to improve the performances on a new domain for several pairs using a domain-specific dataset from a single pair is a very interesting propriety from the mNMT models.

	M2M100		mBART50	
	Baseline	Ours	Baseline	Ours
EN-FI medical	12.93	**14.83**	10.83	10.1
DE-PL medical	12.62	**13.6**	11.1	7.85
FR-IT medical	23.07	**24.62**	10.77	8.58
EN-ES medical	32.38	**35.15**	15.82	17.5
ES-IT medical	25.43	**27.06**	8.40	7.50
ES-FR medical	24.37	**30.64**	19.03	25.6
LT-PL medical	12.49	**13.8**	8.5	7.9
DE-FR medical	18.85	**22.20**	13.3	18.19
LT-PT medical	17.44	**19.26**	*	*

Table 2: Zero shot domain adaptation on medical dataset for other pairs

6.4.2 Comparison of initial pre-trained mNMT models (mBART 50 vs M2M100)

We investigated why mBART50 was more likely to forget other pairs compared to M2M100. First, we have worked with the 418M-parameters version of M2M100. This is not the largest M2M100 version released (and certainly not the most optimized) and this could possibly explain the differences. Then, another hypothesis is the different dataset used during training of both models. Indeed, mBART50 is trained on English-centric data, and M2M100 is not. Non-English centric models are known to achieve higher BLEU especially on low resource data (Fan et al., 2020). Extending this study to domain adaptation, we believe non-English-centric models might be more robust to domain adaptation. We noted that when fine-tuning mBART50 with a bigger learning rate, the first pairs to be forgotten are the non-English ones. Testing this hypothesis on NLLB might be useful.

7 Conclusion and Discussion

In this paper, we propose a study of robust domain adaptation approaches on mNMT models where in-domain data is available only for a single language pair. Best performing approach combines embedding freezing and simple fine-tuning with good hyperparameters. This approach shows good improvements with few in-domain data on all language pairs. The framework effectively avoids overfitting and aggressive forgetting on out-of-domain generic data while quickly adapting to in-domain data. We demonstrate that this could be a solution for incremental adaptation of mNMT models. Finally our work is a call for more research in domain adaptation for multilingual models as it is key for real-world applications.

8 Limitations

This study was limited by hardware issues. We did not have the possibility to fine-tune on M2M100 large version (12B parameters) that requires 64 GB of VRAM.

Testing our results with a larger version of M2M100 might be interesting.

Also, our study focused on two pre-trained multilingual neural machine translation models. However, many others exist and will be released (NLLB Team et al., 2022). We think that our work is generic enough to be applied on other pre-trained models but extensive experiments on these new models should be carried out.

Finally, the work has been realised on English to French data. We showed domain adaptation is possible for languages with English morphology and tested the impact of this training on many different languages morphology (Japanese, English, Russian, ...). Applying domain adaptation training on other morphology languages and on other domains is also an area to investigate.

9 Ethics Statement

The dataset was gathered on OPUS and is largely open-sourced. It was released by (Tiedemann, 2012) and we have downloaded it from OPUS website. We have reviewed the dataset and have not noted any issue with these data. They are very specific to health domain and therefore are not inappropriate. The dataset does not deal with demographic or identity characteristics.

Moreover, these experiments were made using only 2 GPUs and training were relatively short. Given the urgency of addressing climate change, we believe our domain adaptation procedure could help have high-performing mNMT models at small carbon and energy costs. Moreover, SMART framework allows for quicker research of the right hyperparameters, therefore reducing even further the number of experiments and the carbon costs of our method.

10 Acknowledgments

The authors would also like to thank Mr. Baoyang Song, Mr. Laurent Lam, Mr. Laatiri Seif Eddine from BNP Paribas for their valuable comments and suggestions. The work is supported by BNP Paribas.

References

Armen Aghajanyan, Akshat Shrivastava, Anchit Gupta, Naman Goyal, Luke Zettlemoyer, and Sonal Gupta. 2020. Better fine-tuning by reducing representational collapse. *arXiv preprint arXiv:2008.03156*.

Naveen Arivazhagan, Ankur Bapna, Orhan Firat, Roee Aharoni, Melvin Johnson, and Wolfgang Macherey. 2019a. The missing ingredient in zero-shot neural machine translation.

Naveen Arivazhagan, Ankur Bapna, Orhan Firat, Dmitry Lepikhin, Melvin Johnson, Maxim Krikun, Mia Xu Chen, Yuan Cao, George Foster, Colin Cherry, Wolfgang Macherey, Zhifeng Chen, and Yonghui Wu. 2019b. Massively multilingual neural machine translation in the wild: Findings and challenges.

Jimmy Lei Ba, Jamie Ryan Kiros, and Geoffrey E. Hinton. 2016. Layer normalization.

Mauro Cettolo, Nicola Bertoldi, Marcello Federico, Holger Schwenk, Loïc Barrault, and Christophe Servan. 2014. Translation project adaptation for MT-enhanced computer assisted translation. *Machine Translation*, 28:127.

Guanhua Chen, Shuming Ma, Yun Chen, Dongdong Zhang, Jia Pan, Wenping Wang, and Furu Wei. 2022. Towards making the most of cross-lingual transfer for zero-shot neural machine translation. In *Proceedings of the 60th Annual Meeting of the Association for Computational Linguistics (Volume 1: Long Papers)*, pages 142–157, Dublin, Ireland. Association for Computational Linguistics.

Kyunghyun Cho, Bart Van Merriënboer, Caglar Gulcehre, Dzmitry Bahdanau, Fethi Bougares, Holger Schwenk, and Yoshua Bengio. 2014. Learning phrase representations using rnn encoder-decoder for statistical machine translation. *arXiv preprint arXiv:1406.1078*.

Alexis Conneau, Kartikay Khandelwal, Naman Goyal, Vishrav Chaudhary, Guillaume Wenzek, Francisco Guzmán, Edouard Grave, Myle Ott, Luke Zettlemoyer, and Veselin Stoyanov. 2019. Unsupervised cross-lingual representation learning at scale. *arXiv preprint arXiv:1911.02116*.

Alexis Conneau, Guillaume Lample, Marc'Aurelio Ranzato, Ludovic Denoyer, and Hervé Jégou. 2017. Word translation without parallel data. *arXiv preprint arXiv:1710.04087*.

Raj Dabre, Chenhui Chu, and Anoop Kunchukuttan. 2020. A survey of multilingual neural machine translation. *ACM Computing Surveys (CSUR)*, 53(5):1–38.

Jacob Devlin, Ming-Wei Chang, Kenton Lee, and Kristina Toutanova. 2018. Bert: Pre-training of deep bidirectional transformers for language understanding.

Miguel Domingo, Mercedes García-Martínez, Álvaro Peris, Alexandre Helle, Amando Estela, Laurent Bié, Francisco Casacuberta, and Manuel Herranz. 2019. Incremental adaptation of NMT for professional post-editors: A user study. *CoRR*, abs/1906.08996.

Daxiang Dong, Hua Wu, Wei He, Dianhai Yu, and Haifeng Wang. 2015. Multi-task learning for multiple language translation. In *Proceedings of the 53rd Annual Meeting of the Association for Computational Linguistics and the 7th International Joint Conference on Natural Language Processing (Volume 1: Long Papers)*, pages 1723–1732, Beijing, China. Association for Computational Linguistics.

Angela Fan, Shruti Bhosale, Holger Schwenk, Zhiyi Ma, Ahmed El-Kishky, Siddharth Goyal, Mandeep Baines, Onur Celebi, Guillaume Wenzek, Vishrav Chaudhary, Naman Goyal, Tom Birch, Vitaliy Liptchinsky, Sergey Edunov, Edouard Grave, Michael Auli, and Armand Joulin. 2020. Beyond english-centric multilingual machine translation. arXiv.

Orhan Firat, Kyunghyun Cho, and Yoshua Bengio. 2016. Multi-way, multilingual neural machine translation with a shared attention mechanism.

Jiatao Gu, Yong Wang, Kyunghyun Cho, and Victor O. K. Li. 2019. Improved zero-shot neural machine translation via ignoring spurious correlations.

Kaiming He, Xiangyu Zhang, Shaoqing Ren, and Jian Sun. 2015. Deep residual learning for image recognition.

Neil Houlsby, Andrei Giurgiu, Stanislaw Jastrzebski, Bruna Morrone, Quentin de Laroussilhe, Andrea Gesmundo, Mona Attariyan, and Sylvain Gelly. 2019. Parameter-efficient transfer learning for nlp.

Jeremy Howard and Sebastian Ruder. 2018. Universal language model fine-tuning for text classification.

Haoming Jiang, Pengcheng He, Weizhu Chen, Xiaodong Liu, Jianfeng Gao, and Tuo Zhao. 2020. SMART: Robust and efficient fine-tuning for pre-trained natural language models through principled regularized optimization. In *Proceedings of the 58th Annual Meeting of the Association for Computational Linguistics*. Association for Computational Linguistics.

Melvin Johnson, Mike Schuster, Quoc V. Le, Maxim Krikun, Yonghui Wu, Zhifeng Chen, Nikhil Thorat, Fernanda Viégas, Martin Wattenberg, Greg Corrado, Macduff Hughes, and Jeffrey Dean. 2017. Google's multilingual neural machine translation system: Enabling zero-shot translation. *Transactions of the Association for Computational Linguistics*, 5:339–351.

Mike Lewis, Yinhan Liu, Naman Goyal, Marjan Ghazvininejad, Abdelrahman Mohamed, Omer Levy, Ves Stoyanov, and Luke Zettlemoyer. 2019. Bart: Denoising sequence-to-sequence pre-training for natural language generation, translation, and comprehension.

Zhizhong Li and Derek Hoiem. 2016. Learning without forgetting.

Mathis Linger and Mhamed Hajaiej. 2020. Batch clustering for multilingual news streaming. *arXiv preprint arXiv:2004.08123*.

Danni Liu, Jan Niehues, James Cross, Francisco Guzmán, and Xian Li. 2021. Improving zero-shot translation by disentangling positional information.

Yinhan Liu, Myle Ott, Naman Goyal, Jingfei Du, Mandar Joshi, Danqi Chen, Omer Levy, Mike Lewis, Luke Zettlemoyer, and Veselin Stoyanov. 2019. Roberta: A robustly optimized bert pretraining approach.

Minh-Thang Luong, Hieu Pham, and Christopher D Manning. 2015. Effective approaches to attention-based neural machine translation. *arXiv preprint arXiv:1508.04025*.

Michael Mccloskey and Neil J. Cohen. 1989. Catastrophic interference in connectionist networks: The sequential learning problem. *The Psychology of Learning and Motivation*, 24:104–169.

Idriss Mghabbar and Pirashanth Ratnamogan. 2020. Building a multi-domain neural machine translation model using knowledge distillation. *arXiv preprint arXiv:2004.07324*.

NLLB Team, Marta R. Costa-jussà, James Cross, Onur Çelebi, Maha Elbayad, Kenneth Heafield, Kevin Heffernan, Elahe Kalbassi, Janice Lam, Daniel Licht, Jean Maillard, Anna Sun, Skyler Wang, Guillaume Wenzek, Al Youngblood, Bapi Akula, Loic Barrault, Gabriel Mejia Gonzalez, Prangthip Hansanti, John Hoffman, Semarley Jarrett, Kaushik Ram Sadagopan, Dirk Rowe, Shannon Spruit, Chau Tran, Pierre Andrews, Necip Fazil Ayan, Shruti Bhosale, Sergey Edunov, Angela Fan, Cynthia Gao, Vedanuj Goswami, Francisco Guzmán, Philipp Koehn, Alexandre Mourachko, Christophe Ropers, Safiyyah Saleem, Holger Schwenk, and Jeff Wang. 2022. No language left behind: Scaling human-centered machine translation.

Hassan Sajjad, Nadir Durrani, Fahim Dalvi, Yonatan Belinkov, and Stephan Vogel. 2017. Neural machine translation training in a multi-domain scenario. *arXiv preprint arXiv:1708.08712*.

Christophe Servan, Josep Crego, and Jean Senellart. 2016a. Domain specialization: a post-training domain adaptation for neural machine translation. *arXiv preprint arXiv:1612.06141*.

Christophe Servan, Josep Maria Crego, and Jean Senellart. 2016b. Domain specialization: a post-training domain adaptation for neural machine translation. *CoRR*, abs/1612.06141.

Asa Cooper Stickland and Iain Murray. 2019. Bert and pals: Projected attention layers for efficient adaptation in multi-task learning.

Ilya Sutskever, Oriol Vinyals, and Quoc V Le. 2014. Sequence to sequence learning with neural networks. *Advances in neural information processing systems*, 27.

Yuqing Tang, Chau Tran, Xian Li, Peng-Jen Chen, Naman Goyal, Vishrav Chaudhary, Jiatao Gu, and Angela Fan. 2020. Multilingual translation with extensible multilingual pretraining and finetuning.

Jörg Tiedemann. 2012. Parallel data, tools and interfaces in opus. In *Proceedings of the Eight International Conference on Language Resources and Evaluation (LREC'12)*, Istanbul, Turkey. European Language Resources Association (ELRA).

Marlies Van Der Wees, Arianna Bisazza, and Christof Monz. 2017. Dynamic data selection for neural machine translation. *arXiv preprint arXiv:1708.00712*.

Ashish Vaswani, Noam Shazeer, Niki Parmar, Jakob Uszkoreit, Llion Jones, Aidan N. Gomez, Lukasz Kaiser, and Illia Polosukhin. 2017. Attention is all you need.

Shuoheng Yang, Yuxin Wang, and Xiaowen Chu. 2020. A survey of deep learning techniques for neural machine translation.

Jiali Zeng, Jinsong Su, Huating Wen, Yang Liu, Jun Xie, Yongjing Yin, and Jianqiang Zhao. 2018. Multi-domain neural machine translation with word-level domain context discrimination. In *Proceedings of the 2018 Conference on Empirical Methods in Natural Language Processing*, pages 447–457.

Chen Zhu, Yu Cheng, Zhe Gan, Siqi Sun, Tom Goldstein, and Jingjing Liu. 2019. Freelb: Enhanced adversarial training for natural language understanding. *arXiv preprint arXiv:1909.11764*.

Fine-grained Multi-lingual Disentangled Autoencoder for Language-agnostic Representation Learning

Zetian Wu*
Oregon State University
wuzet@oregonstate.edu

Zhongkai Sun
Amazon Alexa AI
zhongkas@amazon.com

Zhengyang Zhao
Amazon Alexa AI
zzhengya@amazon.com

Sixing Lu
Amazon Alexa AI
cynthilu@amazon.com

Chengyuan Ma
Amazon Alexa AI
mchengyu@amazon.com

Chenlei Guo
Amazon Alexa AI
guochenl@amazon.com

Abstract

Encoding both language-specific and language-agnostic information into a single high-dimensional space is a common practice of pre-trained Multi-lingual Language Models (pMLM). Such encoding has been shown to perform effectively on natural language tasks requiring semantics of the whole sentence (e.g., translation). However, its effectiveness appears to be limited on tasks requiring partial information of the utterance (e.g., multi-lingual entity retrieval, template retrieval, and semantic alignment). In this work, a novel Fine-grained Multilingual Disentangled Autoencoder (FMDA) is proposed to disentangle fine-grained semantic information from language-specific information in a multi-lingual setting. FMDA is capable of successfully extracting the disentangled template semantic and residual semantic representations. Experiments conducted on the MASSIVE dataset demonstrate that the disentangled encoding can boost each other during the training, thus consistently outperforming the original pMLM and the strong language disentanglement baseline on monolingual template retrieval and cross-lingual semantic retrieval tasks across multiple languages.

1 Introduction

Pre-trained multilingual language models such as multilingual BERT (mBERT) (Devlin et al., 2019) and XLM-RoBERTa (XLM-R) (Conneau et al., 2020) have been extensively explored and used in academia and industry. These models encode both language-specific information (e.g., grammar, tense, syntax) and language-agnostic information (e.g., semantic, entity, sentiment) into one high-dimensional embedding. However, it has been demonstrated that such encoders perform poorly in some tasks due to a lack of capacity to disentangle fine-grained language-agnostic and language-specific information. (Tiyajamorn et al., 2021; Wieting et al., 2020; Roy et al., 2020; Ahuja et al., 2020; Lin et al., 2021; Asai et al., 2020).

Table 1 presents several application examples where disentangled language-specific or language-agnostic encoder might have better performance. The first example is cross-lingual retrieval, in which a English utterance with the same language-agnostic semantic can be retrieved by a German utterance. Note the semantics in both template and slot can be maintained. The second example is a template retrieval, in which "do you show me doing" can be replaced by another similar meaning template "can you show me how to do" while the slot value changes. The third example is paraphrase retrieval, in which the target utterance keeps slot text but rephrases the template part. Although the applications shown above are retrieval tasks but all of them can be used in query reformulation (Ponnusamy et al., 2020, 2022) and data augmentation (Xu et al., 2021; Kale and Rastogi, 2020; Liu et al., 2021; Gao et al., 2022). For instance, the source utterance "do you show me doing [exercise: backflip]" is a defective sentence with grammar error, and the disentangled encoder is able to retrieve a similar meaning but grammar correct utterance by ignoring the uncommon slot value "backflip".[1] The tasks of cross-lingual retrieval and paraphrase retrieval both are commonly used for data augmentation, especially for languages with data scarcity.

In this work, we proposed a lightweight encoding architecture called Fine-grained Multilingual Disentangled Autoencoder (FMDA) that can disentangle semantic representations at different aspects. The training of proposed encoder adopts reconstruction loss and contrastive learning. The contributions of our proposal are as follows:

*This work is finished during the internship at Amazon Alexa AI

[1]For this example, the golden reformulation for the defective query would be "can you show me how to do backflip". However, directly performing utterance-level retrieval may fail to find the golden reformulation because of data scarcity. Therefore, template-level retrieval is useful here.

Proceedings of the Massively Multilingual Natural Language Understanding Workshop (MMNLU-22), pages 12 - 24
December 7, 2022

Application Tasks	Utterances	Language	Semantic	Template	Slot
cross-lingual retrieval	source: wecke mich um [time: fünf uhr] auf target: wake me up at [time: five am]	different	same	same/similar	same
template retrieval	source: do you show me doing [exercise: backflip] target: can you show me how to do [exercise: yoga]	same	related	same/similar	different
paraphrase retrieval	source: what can be seen inside [object: the basket] target: what does [object: the basket] mainly contain	same	same	different	same

Table 1: Applications of Disentangled Semantic

1. The FMDA is able to extract embedding of:

- language agnostic template representation that contains the semantic information related to the sentence backbone. E.g., in the sentence "Can you play the music Green Light", the template representation aims at encoding the semantic of the "Can you play the music []";
- language agnostic meaning representation which contains both template semantic and residual semantic (e.g., slot name "Green Light") information;
- language-specific non-semantic representation that contains unique language facts;

Visualizations of these fine-grained embedding representations are shown in Figure 4 in Section 4.5.

2. The FMDA designs multiple contrastive learning objectives to improve the performance of the disentanglement learning.

3. Compared with the original pMLM and a language-disentanglement SOTA (Tiyajamorn et al., 2021), FMDA achieves significant improvement on both monolingual template retrieval and cross-lingual meaning retrieval tasks, evaluated on the benchmark MASSIVE (FitzGerald et al., 2022).

4. An ablation study further proves the effectiveness of our model, and a two-stage training experiment has been conducted to further study the effect of the fine-grained semantic.

2 Related Work

Multilingual sentence encoders are widely studied and applied to downstream tasks in recent years. Self-attention networks based multilingual sentence encoders, such as mBERT (Devlin et al., 2019) and XLM-R (Conneau et al., 2020), are pretrained on multilingual corpora in over 100 languages. LaBSE (Feng et al., 2022) encodes text to multilingual sentence embedding by training with 100 million sentence pairs in bilingual corpora of 109+ languages. Libovický et al. (2020) proposes a centered embedding method that subtracts the mean embedding for each language from the sentence embedding, as well as a projection embedding method that projects bilingual using a parallel corpus. MUSE (Chidambaram et al., 2018; Yang et al., 2019) applies a translation based ranking task to one-billion weblab QA pairs to obtain a multilingual universal encoder. Multilingual SBERT (Reimers and Gurevych, 2020) extends pre-trained monolingual SBERT (Reimers and Gurevych, 2019) to the multi-lingual version by mapping translations and original utterances into the same space.

Beyond atomic encoding, some research also focus on disentangling language specific and language-agnostic embeddings. (Chen et al., 2019) learns to disentangle language syntax and semantic information by using aligned paraphrase data to train semantic and use word-order information to train syntax. BGT (Wieting et al., 2019) utilizes a deep variational probabilistic model together with transformers to learn better semantic embeddings in a bi-lingual setting by excluding language-specific information from the information shared across languages. Tiyajamorn et al. (2021) proposes a method for distilling language-agnostic meaning embeddings by removing language-specific information from sentence embeddings generated by off-the-shelf multilingual sentence encoders. Although these works extract both language-specific and language-agnostic embeddings, they are hard to support fine-grained semantic disentanglement.

Based on the work of Tiyajamorn et al. (2021), we further extend the semantic extraction to a fine-grained level. Specifically, our proposed method FMDA is able to extract not only the semantic information of the whole utterance but also part of it, i.e. template/carrier phrase semantic information by learning to disentangle language information at different levels.

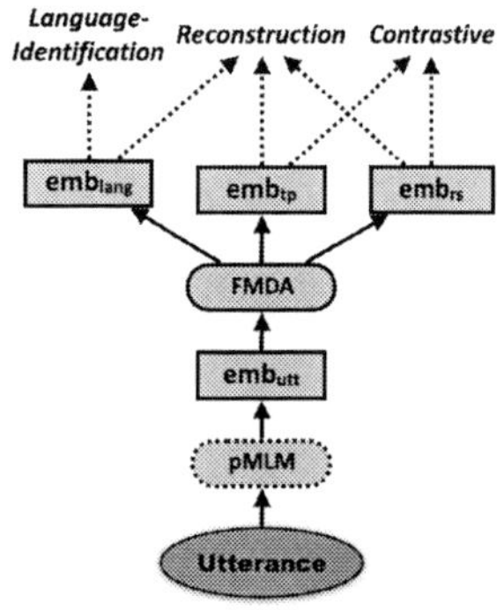

(a) Model Overview

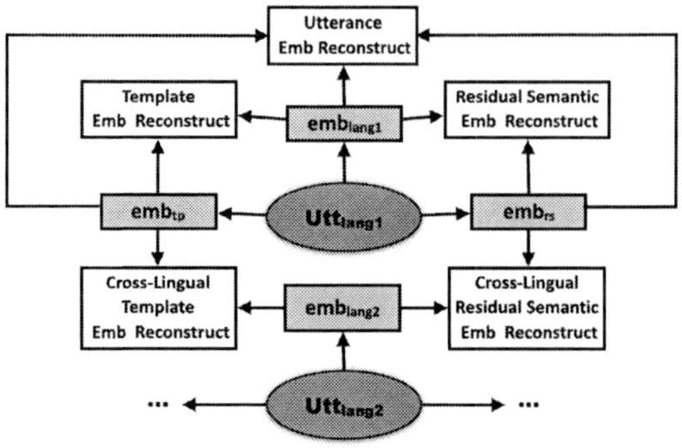

(b) Reconstruct Objective

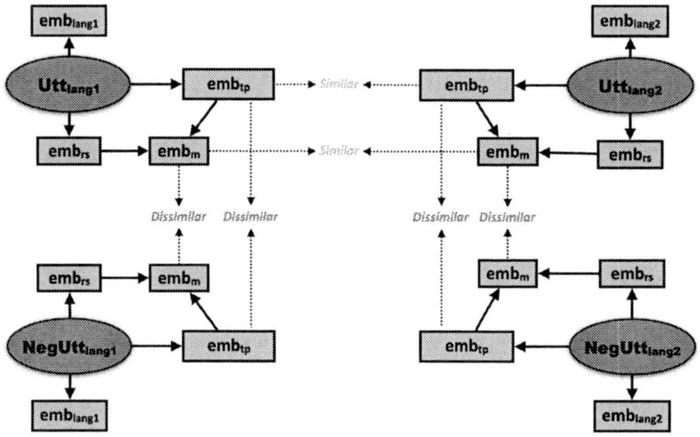

(c) Contrastive Objective

Figure 1: The Overview of our model. (a) The FMDA outputs language-specific non-semantic embedding (emb_{lang}), the template semantic embedding (emb_{tp}), and the residual-semantic embedding (emb_{rs}). Three objectives are applied to train the FMDA: language identification, reconstruction, and contrastive learning. (b) The emb_{tp}, emb_{rs}, and emb_{lang1} are used to reconstruct the original pMLM's utterance embedding; the emb_{lang1} together with The emb_{tp} or emb_{rs} are used for template embedding or residual-semantic embedding reconstruction, respectively; the emb_{lang2} from the utterance in another language but with same meaning can be used with emb_{tp} and emb_{rs} for the cross-lingual reconstruction; (c) Contrastive learning objectives are applied to both emb_{tp} and emb_m (obtained from both emb_{tp} and emb_{rs}).

3 Method

This section describes the details of our proposed Fine-grained Multilingual Disentangled Autoencoder (FMDA). Figure 1(a) demonstrates the overview of the method. FMDA is trained to extract language-specific non-semantic embedding, template semantic embedding, and residual-semantic embedding from a pMLM with three objectives: language identification, embedding reconstruction, and contrastive learning. Figure 1(b) presents the objectives of the embedding reconstruction. Figure 1(c) shows the detailed contrastive leairning design, in which FMDA tries to minimize embedding difference between two paired utterances in different languages and maximize the embedding difference between two non-related utterances in the same language. Each training step takes four utterances as inputs: utterance in language 1, utterance with same-semantic in language 2, and negative utterances with different semantic in language 1 and 2.

3.1 Fine-grained Model Details

Figure 2 demonstrates the detail of our proposed fine-grained model. Given an utterance, a pMLM is first utilized to extract the utterance embedding $emb_{utt} \in R^{L \times D}$, where L represents the utterance length and D represents the embedding dimension. Note that the pMLM's parameters are frozen and the emb_{utt} will remain unchanged during the training and inference. So that the original informa-

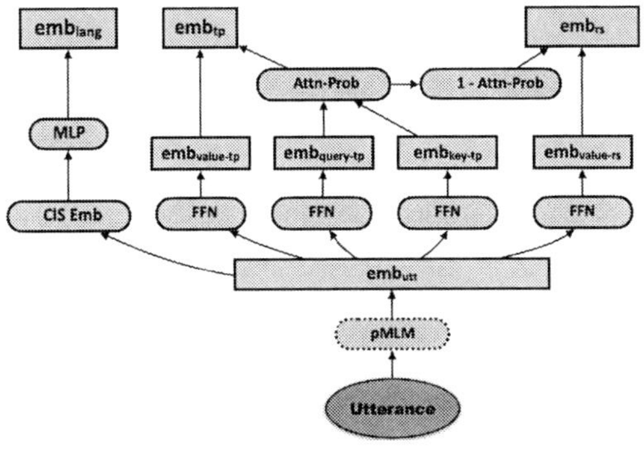

Figure 2: The details of the FMDA. (a) given the utterance embedding emb_{utt} obtained from the pMLM, a MLP layer is first applied to the $[CLS]$ token embedding to generate the emb_{lang}; four FFN layers are applied to the emb_{utt} to generate the template value-embedding ($Value_{tp}$), template query-embedding ($Query_{tp}$), template key-embedding (Key_{tp}), and the residual-semantic value-embedding ($Value_{rs}$). The $Query_{tp}$ and Key_{tp} are first used to generate the attention-probability $Attn - Prob$ for template, therefore the final template embedding can be generated using $Value_{rp}$ and $Attn - Prob$. Besides, the $1 - Attn - Prob$ is also calculated to represent the residual-semantic attention, which can then be used with the $Value_{rs}$ to generate the final emb_{rs}.

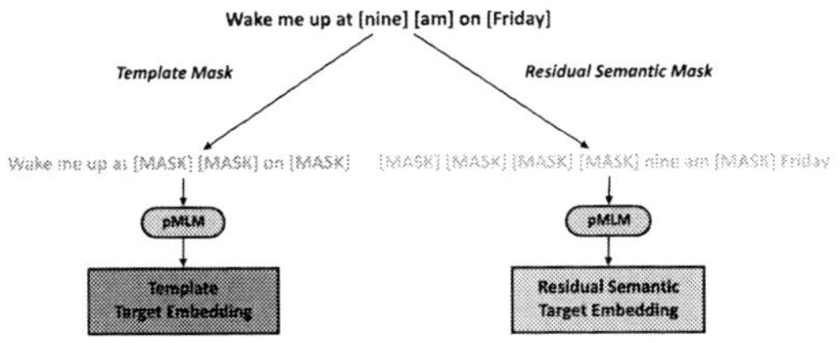

Figure 3: Overview of how to generate the template and residual-semantic reconstruction target embeddings.

tion encoded by the pMLM will be reserved and our method can be lightweight to extract the embeddings of interest. Next, FeedForward-Network (FFN) layers are applied to the emb_{utt} to extract three different embeddings:

Language-Specific Non-semantic Embedding The $[CLS]$ token embedding ($emb_{cls} \in R^{1 \times D}$) of the emb_{utt} is extracted and input to a Multi-Layer-Perceptron (MLP) to obtain the language-specific non-semantic embedding $emb_{lang} \in R^{1 \times D}$.

Language-agnostic Template Semantic embedding In order to encode the semantic information more effectively, an attention-based method is applied to the emb_{utt}. Specifically, three different FFN layers are applied to the emb_{utt} to extract the template value embeddings ($emb_{value-tp} \in R^{L \times D}$), template query-embeddings ($emb_{query-tp} \in R^{L \times D}$), and the template key-embeddings ($emb_{key-tp} \in R^{L \times D}$). The template attention-probability Attn-prob $\in L \times L$ can then be calculated as Softmax($emb_{query-tp} \cdot emb_{key-tp}^T$). The calculated Attn-prob is thereby used with the template value embeddings ($emb_{value-tp}$) to obtain the attention based template embeddings sequence $emb_{tp-seq} \in R^{L \times D}$): $emb_{tp-seq} =$ Attn-prob $\cdot\, emb_{value-tp}$.

Finally, the $[CLS]$ token position of the emb_{tp-seq} is extracted as the final template semantic embedding $emb_{tp} \in R^{1 \times D}$.

Language-agnostic residual-semantic embedding FMDA leverages the Attn-prob learned from the generation of template embedding to generate the embedding for residual semantic. The motivation is to disentangle the template semantic and residual-semantic as much as possible, i.e., the information that template does not pay attention to should be used more to generate the residual-semantic embedding.

Therefore, given the template Attn-prob learned above, we first calculate its opposite (1 - Attn-prob) to obtain the residual-semantic Attn-prob $\in R^{L \times L}$: RS Attn-prob $=$ softmax($1 -$ Attn-prob). Then, the residual-semantic embedding sequence $emb_{rs} \in R^{1 \times D}$ can be obtained.

3.2 Language Identification Objective

To ensure that the extracted language-specific embedding emb_{lang} contains correct language information, following the idea in Tiyajamorn et al. (2021), the emb_{lang} is used for a language-identification objective. Specifically, the emb_{lang} will be input to a MLP layer to achieve the language prediction P, which is used with the true language label L to calculate the language-identification loss $Loss_{lang} =$ CrossEntropy(P, L).

At each training step, the $Loss_{lang}$ is calculated for both of the utterance in language 1 and the utterance with same semantic in language 2.

3.3 Reconstruction Objective

Utterance Embedding Reconstruction As shown in Figure 1(b), for the utterance in language 1 and its emb_{utt} obtained from the pMLM, the $emb_{lang1}, emb_{tp}, emb_{rs}$ learned from FMDA are expected to not lose any information in the original emb_{utt}. Therefore, $emb_{lang1}, emb_{tp}, emb_{rs}$ are used to reconstruct the emb_{utt}. MSE (mean-squared-error) is used as the loss function, thus the $Loss_{rec}^{utt-l1}$ for the utterance in language 1 can be defined as $Loss_{rec}^{utt-l1} = MSE(MLP(emb_{lang1} + emb_{tp} + emb_{rs}), emb_{utt})$

Similarly, the $Loss_{rec}^{utt-l2}$ is also calculated for the utterance with same semantic in language 2.

Template Embedding Reconstruction The emb_{lang} and emb_{tp} are used to reconstruct the template target embedding $emb_{tp}^{Tgt} \in R^{1\times D}$.

Figure 3 demonstrates how to obtain the target template embedding emb_{tp}^{Tgt-L1} together with the residual-semantic target embedding emb_{tp}^{Tgt-L1} for the utterance in language 1. Specifically, given an utterance with token labels indicating if a token belongs to the template or slots, the original utterance is then masked according to the labels and input to the pMLM to obtain the $[CLS]$ position's embeddings as the template target embedding and residual-semantic target embedding.

Both emb_{lang} and emb_{tp} are used to reconstruct the emb_{tp}^{Tgt} as emb_{tp} should contain language-agnostic template-semantic only, while the template target emb_{tp}^{Tgt} contains some language-specific information obtained from the pMLM.

In order to calculate the loss, the sum of the emb_{tp} and emb_{lang1} is obtained as $emb_{tpL1} \in R^{1\times D}$, which is then used to reconstruct the emb_{tp}^{Tgt-L1}. Therefore, the $Loss_{rec}^{cpL1}$ can be defined as:

$$Loss_{rec}^{tpL1} = MSE(emb_{tpL1}, emb_{tp}^{Tgt-L1}) \quad (1)$$

Besides, the emb_{tp} is also combined with emb_{lang2}, which is obtained from the utterance with the same semantic but in language 2, to build the to $emb_{tpL1-CL}$ to reconstruct the template target embedding in language2 (emb_{tp}^{Tgt-L2}). This objective further guarantees that the emb_{tp} contains language-agnostic template semantic information. The cross-lingual template loss $Loss_{rec}^{tpL1-CL}$ can then be defined as:

$$Loss_{rec}^{tpL1-CL} = MSE(emb_{tpL1-CL}, emb_{tp}^{Tgt-L2}) \quad (2)$$

The same process is also applied to the utterance with the same semantic in language 2, to calculate the $Loss_{rec}^{tpL2}$ and $Loss_{rec}^{tpL2-CL}$.

Residual-semantic Embedding Reconstruction Similar reconstruction objectives are applied to the residual-semantic. After obtaining the residual-semantic target embedding emb_{rs}^{Tgt} as illustrated in Figure 3, the $Loss_{rec}^{rsL1}$ and $Loss_{rec}^{rsL1-CL}$ can be calculated for the utterance in language 1 with the same process in the template reconstruction.

Similarly, the functions $Loss_{rec}^{rsL2}$ and $Loss_{rec}^{rsL2-CL}$ are calculated for the utterance with the same semantic in language 2.

3.4 Contrastive Objective

Figure 1 (c) demonstrates the idea of the contrastive learning objective. The contranstive learning is applied to both of the template semantic embeddings and the utterance-level semantic (meaning) embeddings. For the utterances with the same semantic in different languages, their template semantic embeddings and meaning embeddings should be similar; for the utterances in the same language but with different semantics, their template semantic embeddings and meaning embeddings should be different.

Given the inputs as the followings: utterance in language 1, utterance with same-semantic in language 2, and negative utterances with different semantic in language 1 and 2, the contrastive loss for the template semantic $Loss_{con}^{tp}$ can be calculated as:

$$\begin{aligned} Loss_{con}^{cp} = &- \text{Cos-sim}(emb_{tp}^{L1}, emb_{tp}^{L2}) \\ &+ \text{Cos-sim}(emb_{tp}^{L1}, emb_{neg-tp}^{L1}) \quad (3) \\ &+ \text{Cos-sim}(emb_{tp}^{L2}, emb_{neg-tp}^{L2}) \end{aligned}$$

where emb_{tp}^{L1} and emb_{tp}^{L2} come from the utterances with the same semantic but in language 1 and 2; emb_{neg-tp}^{L1} and emb_{neg-tp}^{L2} denote the template semantic embeddings of negative utterances that with different semantic in language 1 and 2.

To conduct the contrastive learning for the meaning embedding, a MLP is first used to generate the meaning embedding $emb_m \in R^{1\times D}$ using the sum of emb_{tp} and emb_{rs}. Then the contrastive loss for the meaning $Loss_{con}^{m}$ can be calculated as:

$$\begin{aligned} Loss_{con}^{m} = &- \text{Cos-sim}(emb_{m}^{L1}, emb_{m}^{L2}) \\ &+ \text{Cos-sim}(emb_{m}^{L1}, emb_{neg-m}^{L1}) \quad (4) \\ &+ \text{Cos-sim}(emb_{m}^{L2}, emb_{neg-m}^{L2}) \end{aligned}$$

Similarly, emb_{m}^{L1} and emb_{m}^{L2} represent the utterance-level semantic meaning of the utterances with the same semantic but from language 1 and 2; emb_{neg-m}^{L1} and emb_{neg-m}^{L2} denote the meaning embeddings from negative utterances that with different semantic from language 1 and 2, respectively.

3.5 Total Training Objective

During the training stage, all of the objectives' loss functions will be optimized together. Therefore, the total loss can be written as:

$$
\begin{aligned}
Loss_{total} = & Loss_{lang}^{L1} + Loss_{lang}^{L2} \\
& + Loss_{rec}^{utt-L1} + Loss_{rec}^{utt-L2} \\
& + Loss_{rec}^{tpL1} + Loss_{rec}^{tpL2} \\
& + Loss_{rec}^{tpL1-CL} + Loss_{rec}^{tpL2-CL} \\
& + Loss_{rec}^{rsL1} + Loss_{rec}^{rsL2} \\
& + Loss_{rec}^{rsL1-CL} + Loss_{rec}^{rsL2-CL} \\
& + Loss_{con}^{tp} + Loss_{con}^{m}
\end{aligned} \tag{5}
$$

4 Experiments

This section describes the experiments conducted on various language pairs using the multilingual natural language understanding dataset MASSIVE (FitzGerald et al., 2022). To evaluate the proposed FMDA, two retrieval tasks introduced in Table 1 are used: (1) cross-lingual semantic retrieval, where the goal is to find the best semantically matching utterance pairs from two languages; and (2) monolinugal template retrieval, where the goal is to find utterances with different slot values but similar template in one language.

4.1 Dataset

Both training and evaluation of our experiment were conducted using the MASSIVE dataset (FitzGerald et al., 2022), which is a cross-lingual corpus that contains virtual assistant utterances across 51 languages. Domains, intents, and slots have been labeled for each utterance.

We chose four languages from MASSIVE - English (EN), German (DE), Spanish (ES) and Japanese (JA) - to form three language pairs (EN-DE, EN-ES, EN-JA) to conduct the experiment. Such selection covers both languages that are similar (e.g. EN-DE) and languages that belongs to distant families (e.g. EN-JA). Our training and evaluation sets were prepared by pre-processing on MASSIVE's train split (containing 11k utterances in each language) and test split (containing 2974 utterances in each language), respectively.

4.2 Setup

XLM-R (base) (Conneau et al., 2020) is used as the backbone encoder to train our proposed FMDA model on three language pairs: EN-DE, EN-ES and EN-JA. As described in Section 3, each training step takes four utterances as inputs: a pair of parallel utterances from language 1 and language 2, and negative utterances with different semantic in language 1 and 2. The following is an example of the training data for EN-DE language pair:

utt_en: Wake me up at nine am on Friday.
utt_de: Weck mich am freitag um neun uhr auf.
neg_utt_en: Quiet.
neg_utt_de: Zeit zu schlafen. (Time to sleep.)

The parallel utterances are directly from the MASSIVE dataset. Negative utterances, on the other hand, are sampled from negative utterance pools. We built a negative utterance pool for each language from the whole training set. For each utterance in the training set, we calculated BLEU scores between it and all other utterances in the same language. We add an utterance into the negative pool if its scores are all smaller than 0.1, which guarantees that utterances in the pool are dissimilar from all other utterances in the training set except itself.

During training, the weights of XLM-R were frozen and only the layers in the FMDA were fine-tuned. The development set from MASSIVE was used to determine the best stop point of training. The other hyperparameters were similar with those used in Tiyajamorn et al., 2021.

To evaluate the output embeddings of our FMDA model, we performed two retrieval tasks as described in Section 4.3.1 and 4.3.2, and compared our results with (a) XLM-R's original $[CLS]$ embedding, and (b) the SOTA language-disentanglement model (Tiyajamorn et al., 2021) trained with our data.

4.3 Results

4.3.1 Cross-lingual Semantic Retrieval Task

We used MASSIVE's test split (containing 2974 sets of parallel utterances) to conduct cross-lingual semantic retrieval evaluation. Given one utterance (query) in the source language, we expect to locate its exact translation from the 2974 candidates in the target language. This was done by calculating cosine similarity between each query and all candidates in the embedding space as the ranking score. The retrieval performance was measure by accu-

Model	Embedding	EN-DE	DE-EN	EN-ES	ES-EN	EN-JA	JA-EN
XLM-R	cls	0.182	0.203	0.194	0.198	0.050	0.037
Tiyajamorn et al. (2021)	meaning	0.550	0.575	0.583	0.602	0.364	0.359
	meaning	**0.594**	**0.605**	**0.645**	**0.650**	**0.400**	**0.380**
Our model	Template(TP)	0.583	0.589	0.630	0.643	0.388	0.371
	Residual(RS)	0.283	0.268	0.369	0.321	0.070	0.075

Table 2: Results of cross-lingual semantic retrieval. **X-Y** in the column headers represents the language pairs for evaluation, where **X** is the source language and **Y** is the target language. The training language pair is the same as the corresponding evluation language pair for each column. The retrieval performance is measured by accuracy@1.

Model	Embedding	$\mathbf{EN}_{EN\_DE}$	$\mathbf{EN}_{EN\_ES}$	$\mathbf{EN}_{EN\_JA}$	DE	ES	JA
XLM-R	cls	0.371	0.371	0.371	0.351	0.427	0.120
Tiyajamorn et al. (2021)	meaning	0.393	0.392	0.299	0.381	0.434	0.267
	meaning	0.427	0.437	0.356	0.387	**0.441**	0.314
Our model	TP	**0.427**	**0.439**	**0.441**	**0.396**	0.436	**0.340**
	RS	0.330	0.348	0.241	0.354	0.420	0.094

Table 3: Results of mono-lingual template retrieval. Column headers show the training and evaluation languages. For example, $\mathbf{EN}_{EN\_DE}$ means the model is trained on EN-DE language pair and evaluated on EN. For language X other than EN, the training language pair is EN-X. The retrieval performance is measured by accuracy@1.

racy@1, i.e. the fraction that the top-1 retrieval matches the target.

Table 2 shows the cross-lingual retrieval result of different models and embeddings. The first row shows the performance of XLM-R's original $[CLS]$ embedding, and the second row shows the performance of the language-agnostic meaning embedding by training the network from Tiyajamorn et al., 2021. The meaning embedding from our FMDA model constantly outperforms both baselines. Diving deeper, we notice that the meaning embedding from Tiyajamorn et al., 2021 may retrieve an utterance with related semantic but of different template and slots; whereas our meaning embedding, which is reconstructed using the fine-grained components, is able to capture the exact translation (as demonstrated by the case in Table 7 of Appendix). The ablation study in Section 4.4 also proves the importance of the fine-grained reconstruction for cross-lingual retrieval.

Comparing the three embedding representations of our model, we find the meaning embedding outperforms the template embedding (TP) as expected, since the former contains more semantic information than the latter (as shown by the case in Table 8 in Appendix). Residual-semantic embedding (RS) in the bottom row has the worst performance because it encodes the least semantic information.

We also notice the differences between language pairs when comparing the columns in Table 2. All embeddings perform much worse on EN-JA than EN-DE/ES, because Japanese belongs to a language family distant from the others. We will further discuss this in Appendix C.

4.3.2 Mono-lingual Template Retrieval Task

To validate the capacity of our model for extracting the carrier phrase/template information from an utterance, we further carried out the mono-lingual template retrieval as the second evaluation task. The evaluation pairs were generated from MASSIVE's test split by manually replacing the slot value of utterances, such that the source and target utterances are from the same language, share the same template, but differs in their slot values (for utterances without a slot labelled, we just discarded them). This pre-processing resulted in about 1.9k evaluation pairs for each language. The following is one of the evaluation pair from EN:

source_utt: I like Senatra songs.

target_utt: I like Taylor Swift songs.

Similar with the cross-lingual semantic retrieval, given a source utterance we expect to find the target utterance from the pool. The performance of the retrieval measured by accuracy@1 is shown in Table 3. As can be seen, the template embedding from our model have consistent better performance than the embedding from the baseline model by Tiyajamorn et al. (2021). Besides, our meaning embedding has the similar performance compared to the template embedding, which means that the meaning embedding is able to contain most of the information from the template embedding.

To make the conclusion of the template retrieval experiment more solid, we prepared an alternative evaluation set, where multiple target utterances were generated from one source, all sharing the same template. Then the retrieval performance measured by mean average precision (MAP) is shown in Table 6 in the Appendix, which matches the observation from Table 3.

4.4 Ablation Study

To understand the importance of each set of losses in the FMDA model, we conducted an ablation study for the cross-lingual retrieval as shown in Table 4. Removing the contrastive loss leads to a significant drop on the retrieval performance, since such loss between a pair of parallel utterances is essential to build up the language alignment.

In addition, among the three sets of reconstruction losses, we find the utterance reconstruction loss $Loss_{rec}^{utt}$ brings significant benefit, while the residual-semantic reconstruction loss $Loss_{rec}^{rs}$ has little function. It needs to noted that when removing all three reconstruction losses, the performance is worse than the original FMDA model, but better than removing $Loss_{rec}^{utt}$ only. This is because in the latter setting, the model with partial reconstruction may lead to a sub-optimal by learning partial information of the utterance. This further proves the effectiveness of the interaction among each reconstruction loss.

Model	EN-DE	DE-EN
XLM-R	0.182	0.203
Tiyajamorn et al. (2021)	0.550	0.575
Our model	**0.594**	0.605
w/o all reconstruction losses	0.580	0.603
w/o utterance reconstruction loss	0.569	0.586
w/o template reconstruction loss	0.583	0.600
w/o residual reconstruction loss	0.584	**0.610**
w/o contrastive loss	0.241	0.234

Table 4: The performance of models with different training settings on cross-lingual semantic retrieval tasks (measured by accuracy@1).

4.5 Visualization

The fine-grained embeddings from our FMDA model are visualized using t-SNE plotting as shown in Figure 4. Figure 4a and 4b show the language embeddings and meaning embeddings of 800 EN-DE utterance pairs, respectively. These embeddings are generated from the FMDA model described in Section 3. Clearly, the language embeddings shows separated language clusters. While the meaning embeddings shows the translation alignment between two languages. In addition, Figure 4b contains multiple clusters, which correspond to different domains/intents in the corpus.

Figure 4c visualizes the template embeddings (generated from the two-stage FMDA model as described in Section 4.6) of 20 English utterances. All of them are from *play-music* intent, but of 4 different templates. The plot shows clearly that our template embedding is efficient in extracting the template information from different sentences.

4.6 Further Exploration with the Two-Stage FMDA

Former experiments demonstrate the effectiveness of fine-grained decomposition and reconstruction of embedding representations using our proposed FMDA model, and its benefit for different applications. However, the training of the FDMA involves multiple different loss functions, which may affect the optimization of each component. Therefore, we would like to investigate if training different components of FDMA in separating steps can lead to better embedding representations.

Here we conducted a two-stage training procedure to obtain better template representation. For the first stage, we focused on template encoder in FMDA and template related loss terms, i.e. the template reconstruction loss and the template contrastive loss. Mono-lingual template pairs data, in the same format as that described in Section 4.3.2, were built as positive pairs for the training. Therefore, the template encoder in FMDA can be better learned on this pure template data. In the second stage, the template encoder in the FMDA was frozen and all other losses except template-related ones were used together on the dual-lingual pair training data described in section 4.

Results of the model trained in two-stage setting are shown in Tabel 5. For mono-lingual template retrieval tasks (columns **EN** and **DE**), the template embedding (TP) obtained through two-stage training is far better than that from the original FMDA model. The meaning embedding also benefits from the boost of TP. For cross-lingual semantic retrieval tasks (columns **EN-DE** and **DE-EN**), although the performance of two-stage TP is low (since the template encoder hasn't been trained with dual-lingual pair data in the two-stage setting), the performance

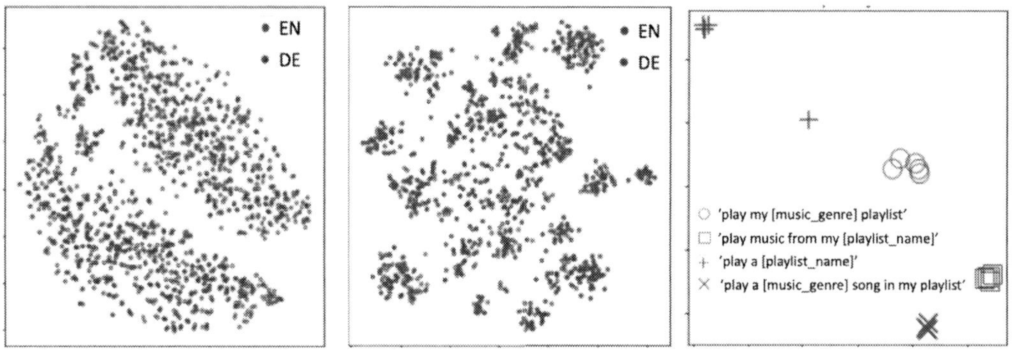

(a) Language embedding (b) Meaning embedding (c) Template embedding

Figure 4: Visualisation of fine-grained embeddings from our FMDA model.

of the meaning embedding is similar with that from the original FMDA model.

This experiment demonstrates that the two-stage based FMDA is able to learn a much better template embedding while the meaning embedding still effectively encodes the whole semantic.

Method	Embedding	EN-DE	DE-EN	EN	DE
All-together	meaning	0.594	**0.605**	0.427	0.387
	TP	0.583	0.589	0.427	0.396
Two-stage	meaning	**0.595**	0.601	0.584	0.566
	TP	0.241	0.251	**0.811**	**0.802**

Table 5: Comparison of all-together training and two-stage training. The numbers of all-together training are from Table 2 and Table 3.

5 Conclusion

In this paper, we introduced FMDA, a lightweight encoding architecture that is able to disentangle fine-grained semantic information from language-specific information in a multilingual setting. Compared with previous works, the FMDA distils 1) language embedding emb_{lang} to encode the language-specific information, 2) template embedding emb_{tp} to encode the the backbone template of the sentence, and 3) the residual embedding emb_{rs} to encode the residual information such as slot. Such fine-grained representations allow retrieval applications at different levels under the NLU setting.

Two retrieval tasks conducted on the MASSIVE dataset demonstrate that FMDA's meaning embedding achieves the best performance on the cross-lingual semantic retrieval task and FMDA's template embedding achieves the best performance on the mono-lingual template retrieval task. Both constantly outperform the SOTA language-disentanglement baseline across multiple languages.

References

Aman Ahuja, Nikhil Rao, Sumeet Katariya, Karthik Subbian, and Chandan K Reddy. 2020. Language-agnostic representation learning for product search on e-commerce platforms. In *Proceedings of the 13th International Conference on Web Search and Data Mining*, pages 7–15.

Akari Asai, Jungo Kasai, Jonathan H Clark, Kenton Lee, Eunsol Choi, and Hannaneh Hajishirzi. 2020. Xor qa: Cross-lingual open-retrieval question answering. *arXiv preprint arXiv:2010.11856*.

Mingda Chen, Qingming Tang, Sam Wiseman, and Kevin Gimpel. 2019. A multi-task approach for disentangling syntax and semantics in sentence representations. *arXiv preprint arXiv:1904.01173*.

Muthuraman Chidambaram, Yinfei Yang, Daniel Cer, Steve Yuan, Yun-Hsuan Sung, Brian Strope, and Ray Kurzweil. 2018. Learning cross-lingual sentence representations via a multi-task dual-encoder model. *arXiv preprint arXiv:1810.12836*.

Alexis Conneau, Kartikay Khandelwal, Naman Goyal, Vishrav Chaudhary, Guillaume Wenzek, Francisco Guzmán, Edouard Grave, Myle Ott, Luke Zettlemoyer, and Veselin Stoyanov. 2020. Unsupervised cross-lingual representation learning at scale. In *Proceedings of the 58th Annual Meeting of the Association for Computational Linguistics*, pages 8440–8451, Online. Association for Computational Linguistics.

Jacob Devlin, Ming-Wei Chang, Kenton Lee, and Kristina Toutanova. 2019. BERT: Pre-training of deep bidirectional transformers for language understanding. In *Proceedings of the 2019 Conference of the North American Chapter of the Association for Computational Linguistics: Human Language Technologies, Volume 1 (Long and Short Papers)*, pages 4171–4186, Minneapolis, Minnesota. Association for Computational Linguistics.

Fangxiaoyu Feng, Yinfei Yang, Daniel Cer, Naveen Arivazhagan, and Wei Wang. 2022. Language-agnostic BERT sentence embedding. In *Proceedings of the 60th Annual Meeting of the Association for Computational Linguistics (Volume 1: Long Papers)*, pages 878–891, Dublin, Ireland. Association for Computational Linguistics.

Jack FitzGerald, Christopher Hench, Charith Peris, Scott Mackie, Kay Rottmann, Ana Sanchez, Aaron Nash, Liam Urbach, Vishesh Kakarala, Richa Singh, Swetha Ranganath, Laurie Crist, Misha Britan, Wouter Leeuwis, Gokhan Tur, and Prem Natarajan. 2022. Massive: A 1m-example multilingual natural language understanding dataset with 51 typologically-diverse languages.

Yifan Gao, Qingyu Yin, Zheng Li, Rui Meng, Tong Zhao, Bing Yin, Irwin King, and Michael Lyu. 2022. Retrieval-augmented multilingual keyphrase generation with retriever-generator iterative training. In *Findings of the Association for Computational Linguistics: NAACL 2022*, pages 1233–1246. Association for Computational Linguistics.

Mihir Kale and Abhinav Rastogi. 2020. Template guided text generation for task-oriented dialogue. *arXiv preprint arXiv:2004.15006*.

Jindřich Libovický, Rudolf Rosa, and Alexander Fraser. 2020. On the language neutrality of pre-trained multilingual representations. In *Findings of the Association for Computational Linguistics: EMNLP 2020*, pages 1663–1674, Online. Association for Computational Linguistics.

Bill Yuchen Lin, Seyeon Lee, Xiaoyang Qiao, and Xiang Ren. 2021. Common sense beyond english: Evaluating and improving multilingual language models for commonsense reasoning. *arXiv preprint arXiv:2106.06937*.

Linlin Liu, Bosheng Ding, Lidong Bing, Shafiq Joty, Luo Si, and Chunyan Miao. 2021. MulDA: A multilingual data augmentation framework for low-resource cross-lingual NER. In *Proceedings of the 59th Annual Meeting of the Association for Computational Linguistics and the 11th International Joint Conference on Natural Language Processing (Volume 1: Long Papers)*, pages 5834–5846. Association for Computational Linguistics.

Pragaash Ponnusamy, Alireza Roshan Ghias, Chenlei Guo, and Ruhi Sarikaya. 2020. Feedback-based self-learning in large-scale conversational ai agents. In *Proceedings of the AAAI conference on artificial intelligence*, volume 34, pages 13180–13187.

Pragaash Ponnusamy, Clint Solomon Mathialagan, Gustavo Aguilar, Chengyuan Ma, and Chenlei Guo. 2022. Self-aware feedback-based self-learning in large-scale conversational AI. In *Proceedings of the 2022 Conference of the North American Chapter of the Association for Computational Linguistics: Human Language Technologies: Industry Track*, pages 324–333.

Nils Reimers and Iryna Gurevych. 2019. Sentence-bert: Sentence embeddings using siamese bert-networks. *arXiv preprint arXiv:1908.10084*.

Nils Reimers and Iryna Gurevych. 2020. Making monolingual sentence embeddings multilingual using knowledge distillation. *arXiv preprint arXiv:2004.09813*.

Uma Roy, Noah Constant, Rami Al-Rfou, Aditya Barua, Aaron Phillips, and Yinfei Yang. 2020. Lareqa: Language-agnostic answer retrieval from a multilingual pool. *arXiv preprint arXiv:2004.05484*.

Nattapong Tiyajamorn, Tomoyuki Kajiwara, Yuki Arase, and Makoto Onizuka. 2021. Language-agnostic representation from multilingual sentence encoders for cross-lingual similarity estimation. In *Proceedings of the 2021 Conference on Empirical Methods in Natural Language Processing*, pages

7764–7774, Online and Punta Cana, Dominican Republic. Association for Computational Linguistics.

John Wieting, Graham Neubig, and Taylor Berg-Kirkpatrick. 2019. A bilingual generative transformer for semantic sentence embedding. *arXiv preprint arXiv:1911.03895*.

John Wieting, Graham Neubig, and Taylor Berg-Kirkpatrick. 2020. A bilingual generative transformer for semantic sentence embedding. In *Proceedings of the 2020 Conference on Empirical Methods in Natural Language Processing (EMNLP)*, pages 1581–1594, Online. Association for Computational Linguistics.

Xinnuo Xu, Guoyin Wang, Young-Bum Kim, and Sungjin Lee. 2021. Augnlg: Few-shot natural language generation using self-trained data augmentation. *arXiv preprint arXiv:2106.05589*.

Yinfei Yang, Daniel Cer, Amin Ahmad, Mandy Guo, Jax Law, Noah Constant, Gustavo Hernandez Abrego, Steve Yuan, Chris Tar, Yun-Hsuan Sung, et al. 2019. Multilingual universal sentence encoder for semantic retrieval. *arXiv preprint arXiv:1907.04307*.

A Multi-target Template Retrieval

In Section 4.3.2, we introduced the mono-lingual template retrieval experiment, which was evaluated using source-target utterance pairs generated by slot replacement. In order to make the conclusion of the experiment more solid, here we prepared an alternative evaluation set, where five target utterances were generated from each source, all of them sharing the same template but with different slot values. Then, mean average precision (MAP) was used to evaluate whether all of the ground-truth targets can be retrieved at high rank. The result is shown in Table 6, which is consistent with the observations from Table 3.

B Case Study

For better understanding of the cross-lingual semantic retrieval results (Section 4.3.1), we pulled out some examples from the EN-DE retrieval experiment to demonstrate the different behaviors of different embeddings.

Table 7 shows a case where the result from our model is different from that of the baseline model (Tiyajamorn et al. (2021)). The meaning embedding from our FMDA model is able to retrieve the correct target, which is the exact translation of the source query. Whereas the embedding from the baseline model retrieves a wrong answer, which has the same intent as the source but differs in template and slot. This is because the semantic representation from our FMDA model is reconstructed from finer grains (template and residual-semantic/slots) and is able to capture detailed information in the sentence more accurately.

Table 8 shows an example where the meaning embedding from our FMDA model captures the correct target while the template embedding from the same model retrieves a wrong one – though its template is same as the query, the slot doesn't match. This is as expected, since the residual-semantic information (slot) is decoupled from the template embedding.

In conclusion, our template embeddings are able to capture necessary template information an utterance, while our meaning embeddings are able to cover the whole semantic of the utterance.

C Extend FMDA to a Unified Multilingual Model

In the former experiments, the models were all trained on dual-lingual pairs, e.g. EN-DE, EN-ES. To further validate if the model can benefit from training multiple languages together, a multi-lingual training experiment is conducted and analyzed. The model architecture remains the same, while the input is not only a dual-lingual pair, but multiple dual-lingual pairs together, i.e., the model is trained on the mixture of EN-DE, EN-ES and EN-JA pairs in one epoch.

Table 9 demonstrates the results of the multi-language pairs training. First, for languages like English (EN), German (DE), and Spanish (ES), the performance of the multi-lingual training model is actually worse than the dual-lingual training model. However, the multi-lingual model performs better on Japanese (JA). Second, for all languages, the multi-lingual training model still outperforms the baseline model from Tiyajamorn et al. (2021).

We argue that the reason of this results is: languages from similar families like EN, DE, and ES have been well learned in the original XLM-R and may have more in common. However, language like JA is a single-family language which does not share common scripts nor in the same genre with others, and is not well-studied in the original XLM-R. Therefore, the performance of the EN, DE, and ES part in the model trained under the multi-lingual setting is affected by JA so that to be worse , while JA, which is insufficient leanred in the XLM-R, can benefit more from other well-learned languages.

Model	Embedding	EN_{EN_DE}	EN_{EN_ES}	EN_{EN_JA}	DE	ES	JA
XLM-R	cls	0.370	0.370	0.370	0.355	0.379	0.117
Tiyajamorn et al. (2021)	meaning	0.397	0.386	0.310	0.386	0.380	0.283
Our model	meaning	0.421	0.432	0.360	**0.409**	0.423	0.350
	TP	**0.421**	**0.434**	**0.395**	0.405	**0.425**	**0.375**
	RS	0.328	0.344	0.234	0.360	0.379	0.088

Table 6: Results of mono-lingual template retrieval under multi-target retrieval setup. The retrieval performance is measured by mean average precision (MAP). Other settings are the same as Table 3.

Query	wake me up at five am this week
Top-1 retrieval	Our model (meaning): wecke mich in dieser woche um fünf uhr auf (wake me up at five am this week)
	Tiyajamorn et al. (2021) model (meaning): ich muss morgen um zehn uhr aufstehen (i need to get up at ten tomorrow)

Table 7: An example from the EN-DE cross-lingual retrieval experiment, for which our meaning embedding retrieved the correct target, whereas the embedding from Tiyajamorn et al. (2021) retrieved a wrong answer.

Query	what's the time in sweden
Top-1 retrieval	Our model (meaning): wie spät ist es in schweden (what's the time in sweden)
	Our model (template): welche uhrzeit ist es in einer stadt (what time is it in a city)

Table 8: An example from the EN-DE cross-lingual retrieval experiment, for which our meaning embedding retrieved the correct target, whereas our template embedding retrieved a wrong answer.

Eval. Task	Tiyajamorn et al. (2021)	Dual-lingual Training	Multi-lingual Training
DE-EN	0.575	**0.605**	0.572
ES-EN	0.602	**0.650**	0.617
JA-EN*	0.359	0.380	**0.397**
EN	0.361	**0.436**	0.429
DE	0.381	**0.396**	0.395
ES	0.434	**0.436**	0.433
JA*	0.267	0.340	**0.341**

Table 9: Comparison of dual-lingual training with multi-lingual training. Results shown are the performance of cross-lingual semantic retrieval (top rows) and mono-lingual template retrieval (bottom rows) respectively, measured by accuracy@1. For dual-lingual training, the model is trained on one language pair and evaluated on the corresponding language (pair), as described in Table 2 and Table 3. For multi-lingual training, a unified model is trained using data from multiple language pairs. Specifically, JA data is not included in the training set except for the rows marked by *.

Evaluating Byte and Wordpiece Level Models for Massively Multilingual Semantic Parsing

Massimo Nicosia and Francesco Piccinno
Google Research, Zürich
{massimon,piccinno}@google.com

Abstract

Token free approaches have been successfully applied to a series of word and span level tasks. In this work, we compare a byte-level (ByT5) and a wordpiece based (mT5) sequence to sequence model on the 51 languages of the MASSIVE multilingual semantic parsing dataset. We examine multiple experimental settings: (i) zero-shot, (ii) full gold data and (iii) zero-shot with synthetic data. By leveraging a state-of-the-art label projection method for machine translated examples, we are able to reduce the gap in exact match accuracy to only 5 points with respect to a model trained on gold data from all the languages. We additionally provide insights on the cross-lingual transfer of ByT5 and show how the model compares with respect to mT5 across all parameter sizes.

1 Introduction

Semantic parsers map natural languages utterances into logical forms (LFs). In the context of conversational agents (Artzi and Zettlemoyer, 2011), robotics (Dukes, 2014) or question answering systems (Berant et al., 2013), task-oriented semantic parsers map user queries (e.g. *"set an 8 am alarm"*) to machine readable LFs (e.g. `[IN:CREATE_ALARM [SL:TIME 8 am ]]`), in the form of structured interpretations that can be understood and executed by downstream components. Learning parsers requires training data in the form of <utterance, LF> pairs. Such data is costly to obtain especially at large scale (Berant et al., 2013), since expert annotators have to derive the correct LFs given an input utterance. This problem is exacerbated in a multilingual setting, where the availability of annotators, especially for non top-tier languages, is scarce and therefore even more expensive.

With the release of MASSIVE (FitzGerald et al., 2022), the research community has now access to a massively multilingual semantic parsing dataset that can be used to evaluate large language models fine-tuned on the task and to study cross-lingual transfer for numerous languages.

On the multilinguality front, token-free models with byte or character based vocabularies have gained strength given their competitiveness with respect to traditional subword-based pretrained language models. Models such as ByT5 (Xu et al., 2020), Canine (Clark et al., 2022) and the Charformer (Tay et al., 2022) have been applied to popular multilingual benchmarks obtaining state-of-the-art results.

In this paper, we perform the first in-depth evaluation of a token-free model in the context of multilingual semantic parsing. We compare the ByT5 and mT5 (Xue et al., 2021) models across different parameter sizes and data regime settings. In addition to that, we build a map of the cross-lingual transfer for all the languages in MASSIVE. Lastly, we show that with the use of machine translated synthetic data the accuracy of a state-of-the-art multilingual parser can be just 5 points lower than the same parser trained with all the available multilingual supervision. To incentivize research on synthetic data augmentation approaches, we release the MASSIVE English training utterances translated to 50 languages.[1]

2 The MASSIVE Dataset

MASSIVE (FitzGerald et al., 2022) is a semantic parsing dataset covering 51 languages, 18 domains, 60 intents and 55 slots. The dataset was created by professional translators starting from the English SLURP dataset (Bastianelli et al., 2020). A significant portion of the translations have been localized too, following the recent trend in multilingual benchmarks of replacing western-centric

[1] We release the translations in 50 languages of the MASSIVE English training examples obtained with an in-house translation system at https://goo.gle/massive-translations

Proceedings of the Massively Multilingual Natural Language Understanding Workshop (MMNLU-22), pages 25 - 34
December 7, 2022

entities with entities that are more relevant for the target languages (Lin et al., 2021; Ding et al., 2022; Majewska et al., 2022).

2.1 Pre and Post Processing

The annotated instances in the MASSIVE dataset come in the following format:

```
intent: alarm_set
annot_utt: despiértame a las [time :
    ↪ nueve de la mañana] el [date :
    ↪ viernes]
```

To shorten the target output and save the model from generating and potentially hallucinating unnecessary words, we map the former to the following format taken from MTOP (Li et al., 2021):

```
[IN:ALARM_SET [SL:TIME nueve de la mañ
    ↪ ana ] [SL:DATE viernes ] ]
```

For evaluation, we use a simple inverse post-processing step based on string matching to convert the model outputs back to MASSIVE format.

2.2 Synthetic Data with Translate-and-Fill

A common approach to create multilingual synthetic data from available examples is to use machine translation (Moradshahi et al., 2020; Sherborne et al., 2020). Utterances are translated and LF annotations are projected using word aligners and noise reduction heuristics. We instead adopt the approach from Nicosia et al. (2021), Translate-and-Fill (TAF), a label projection method in which a filler model reconstructs the full LF starting from an utterance and its LF signature.

We train an mT5-xxl filler model on English instances and then directly generate the LFs of translated examples in a zero-shot fashion. Since the slot order between English and translated utterances may differ, we canonicalize the generated synthetic interpretations reordering the slots as they would occur in the translations. We have also noticed in the filler output that for some languages the slot boundaries may fall inside words. For languages with white space tokenization, we move slot boundaries to word boundaries if needed.

As an example, given an input utterance *"despiértame a las nueve el viernes"* and `[IN:ALARM_SET [SL:DATE el vier ] [SL:TIME nueve ] ]` as LF, the process looks as follows. First the arguments are reordered according to the order of appearance in the original sentence: `[IN:ALARM_SET [SL:TIME nueve ] [SL:DATE vier ] ]`. Then slot boundaries that fall within words are extended, correcting the prediction for the second argument from `[SL:DATE vier ]` to `[SL:DATE viernes ]`.

3 Experiments

We use MASSIVE as a test bed for two model families, ByT5 and mT5, evaluating them at all sizes in three different data settings. We report *Intent Accuracy* (IA) and *Exact Match* (EM) accuracy. We do not perform any hyper-parameter tuning: we train for 30K steps with a fixed learning rate of 0.0001 and a batch size of 128 for all models but xxl, for which batch size was reduced to 32. We run fine tuning on Cloud TPU v3 with an input/-target length of 1024/512 for ByT5 and 512/512 for mT5. To minimize compute, all the reported results are from single runs. We experiment with three different settings, summarized below:

1. **Zero-shot setting.** Training is performed on English data only, and the model selection is done on the English development set. Results are reported in Table 1.

2. **Gold-data setting.** Training is performed on all the MASSIVE data, that includes 51 languages. Model selection is performed averaging the accuracy on the multilingual development sets. Results are reported Table 2.

3. **Synthetic data setting (TAF).** Training is performed on English and multilingual data that is synthetically generated via TAF. Results are reported in Table 3. Our entry based on this approach ranked 1st in the Zero-Shot Task of the MMNLU-22 Multilingual Semantic Parsing competition organized by Amazon and co-located with EMNLP 2022.[2]

We can see a pattern that is common to all the experiments: at smaller sizes, ByT5 has much better EM accuracy then the corresponding mT5 models. As stated in Xu et al. (2020), this may be explained by the fact that at these sizes less than 0.3% of ByT5 parameters are locked in embedding tables and a larger amount of dense parameters is updated during training. mT5 parameters are instead dominated by the embedding tables, which are updated less often than the dense layers. In addition to that, ByT5-large is worse than ByT5-base at span labeling, which is a word level task. Both our observations confirm the findings in Xu et al. (2020).

[2] https://mmnlu-22.github.io

Model	IA	EM
ByT5-small	49.26	20.36
ByT5-base	64.3	33.47
ByT5-large	66.53	28.43
ByT5-xl	80.96	41.7
ByT5-xxl	81.73	38.28
mT5-small	51.75	17.59
mT5-base	55.91	17.73
mT5-large	67.23	25.14
mT5-xl	79.97	45.60
mT5-xxl	**82.44**	**50.21**

Table 1: Zero-shot *T5 parsers performance when training on English only.

Model	IA	EM
ByT5-small	85.59	66.60
ByT5-base	85.93	67.54
ByT5-large	84.02	62.92
ByT5-xl	87.01	68.29
ByT5-xxl	**87.27**	**68.66**
mT5-small	73.29	46.65
mT5-base	82.03	58.24
mT5-large	85.58	64.13
mT5-xl	87.24	68.47
mT5-xxl	86.79	63.33

Table 2: *T5 parsers performance when training on all the available gold data.

In the **synthetic data setting** (Table 3), IA almost matches the IA of models from the gold data setting. If we consider EM accuracy, we are only 5% points behind the upper bound performance of the multilingually supervised -xxl models (see Table 2). This indicates that synthetic data augmentation is a viable approach for the i18n of semantic parsers. Please refer to Table 9 in the appendix for results on individual languages.

4 Additional Experiments and Results

In zero-shot evaluations, English is the most studied language given the availability of labeled data. Recent work has shown that this language may not be the best at cross-lingual transfer (Turc et al., 2021). Since MASSIVE provides training and test data for all its languages, we can evaluate the zero-shot performance of each language. We train 51 ByT5-base model for a fixed number of steps

Model	IA	EM
ByT5-small	83.32	59.32
ByT5-base	84.59	61.24
ByT5-large	82.82	58.09
ByT5-xl	85.90	62.98
ByT5-xxl	86.48	**64.18**
mT5-small	73.64	43.19
mT5-base	80.79	51.76
mT5-large	83.99	57.43
mT5-xl	86.07	62.33
mT5-xxl	**86.69**	62.49

Table 3: *T5 parsers performance when training on English and synthetic TAF data.

(1k steps, 128 batch size) and collect the results on the development sets in Figure 2. By summing the EMs on rows we can understand how much a fine-tuning language (*donor*) improves the others. If we sum over columns, we can see how much transfer a target language (*receiver*) gets from the others. We report some statistics about best/worst donor/receiver languages in Table 4. Interestingly, English is not among the top donors, while it is the one that is being improved the most by other languages. We speculate that the better English LM representations may already have an intrinsic notion of semantic concepts that are then quickly individuated if supervision for such concepts is provided in other languages. From Figure 2, we see that some languages (am, sw, km, cy) clearly need annotated data. We hope that this map could help prioritize data collection efforts.

MASSIVE examples contain an interesting piece of metadata that indicates if an utterance has been translated and localized (i.e. original entities have been substituted with entities more culturally relevant for the target language), or translated only. We split the test sets in two parts according to this information and report in Figure 1 the EM accuracies of the same mT5-xxl model. We examine the three data settings studied in this paper. Accuracies on *localized* utterances are consistently lower. The performance difference in the synthetic data setting is relatively small but it still suggests that creating synthetic examples with entities that are *local* to the target language may improve the robustness of the parser.

In the appendix, we report the accuracy for each individual intent on the union of the test set ex-

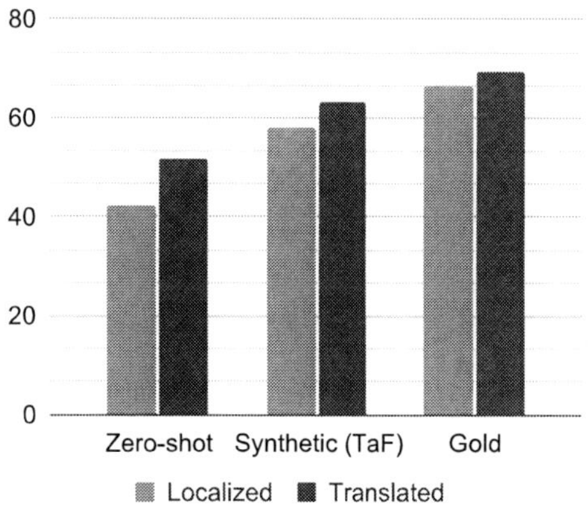

Figure 1: Differences in EM for an mT5-xxl model evaluated on queries of the test set that have been both translated and localized, vs only translated.

	Best to worst
Donor	fr, de, es, nl, pl, ···, mn, am, sw, km, cy
Receiver	en, de, pt, fr, sv, ···, zh, am, mn, sw, cy

Table 4: Top-5 Best/worst donor/receiver.

amples from all languages (Table 8). In Table 5, we report the 6 intents with the lowest accuracy. Most examples belong to the `GENERAL_QUIRKY` intent. The latter is likely a bucket intent covering all the utterances that are generic or out-of-domain (we could not find an exhaustive description of this intent in the SLURP dataset(Bastianelli et al., 2020)). The common parser mistake is to classify such queries as belonging to a more specific intent that can plausibly be associated with that query.

Finally, we compare our NMT translations of the training set with the corresponding gold translations produced by professional translators. We summarize the most interesting information in Table 6 (full comparison in Table 7 included in the appendix). Indic languages (*_IN and bn_BD) have an higher average match than other languages. This may suggest that translations in these languages are more unambiguous or that translators may have relied on a MT during the translation task.

Intent	IA	Support
GENERAL_GREET	19.6	51
MUSIC_SETTINGS	27.1	306
AUDIO_VOLUME_OTHER	54.9	306
GENERAL_QUIRKY	55.6	8619
IOT_HUE_LIGHTON	61.4	153
MUSIC_DISLIKENESS	74.5	204

Table 5: IA of the ByT5-xxl+TaF model for the lowest scoring intents (considering all languages).

Language sets	Avg Match (%)
All languages	21.3
All but Indic languages	17.3
Indic languages	50.8

Table 6: Percentages of NMT translations matching human translations in MASSIVE training set.

5 Related Work

Multilingual models are architecturally similar to monolingual transformer-based models but they are pretrained on multilingual corpora. These models include XLM (Lample and Conneau, 2019), XLM-R (Conneau et al., 2020) and mT5 (Xue et al., 2021), the multilingual version of T5 (Raffel et al., 2020). They all use a subword vocabulary, a choice that may result in poor performance for languages with limited amount of data (Wang et al., 2021). Token-free models such as ByT5 (Xu et al., 2020), Canine (Clark et al., 2022) and Charformer (Tay et al., 2022) were designed to avoid this issue and have been applied to popular multilingual benchmarks obtaining state-of-the-art results. In this work, we compare the multilinguality and the generative capabilities of mT5 and ByT5 in a massively multilingual semantic parsing task.

Data augmentation is the process of creating synthetic labeled data from available annotated examples. One approach in the multilinguality space is to translate annotated data in one language, e.g. English, to other languages. Neural machine translation is a strong baseline as it has been shown in recent cross-lingual evaluation benchmarks (Hu et al., 2020; Ladhak et al., 2020). While translation works quite well for classification tasks where the label is at instance level, sequence tagging or parsing tasks require an annotation projection step because labels are at token level. Translate-and-align methods use bilingual word aligners, statistical (Brown et al., 1993; Vogel et al., 1996; Och and Ney, 2000, 2003), and neural

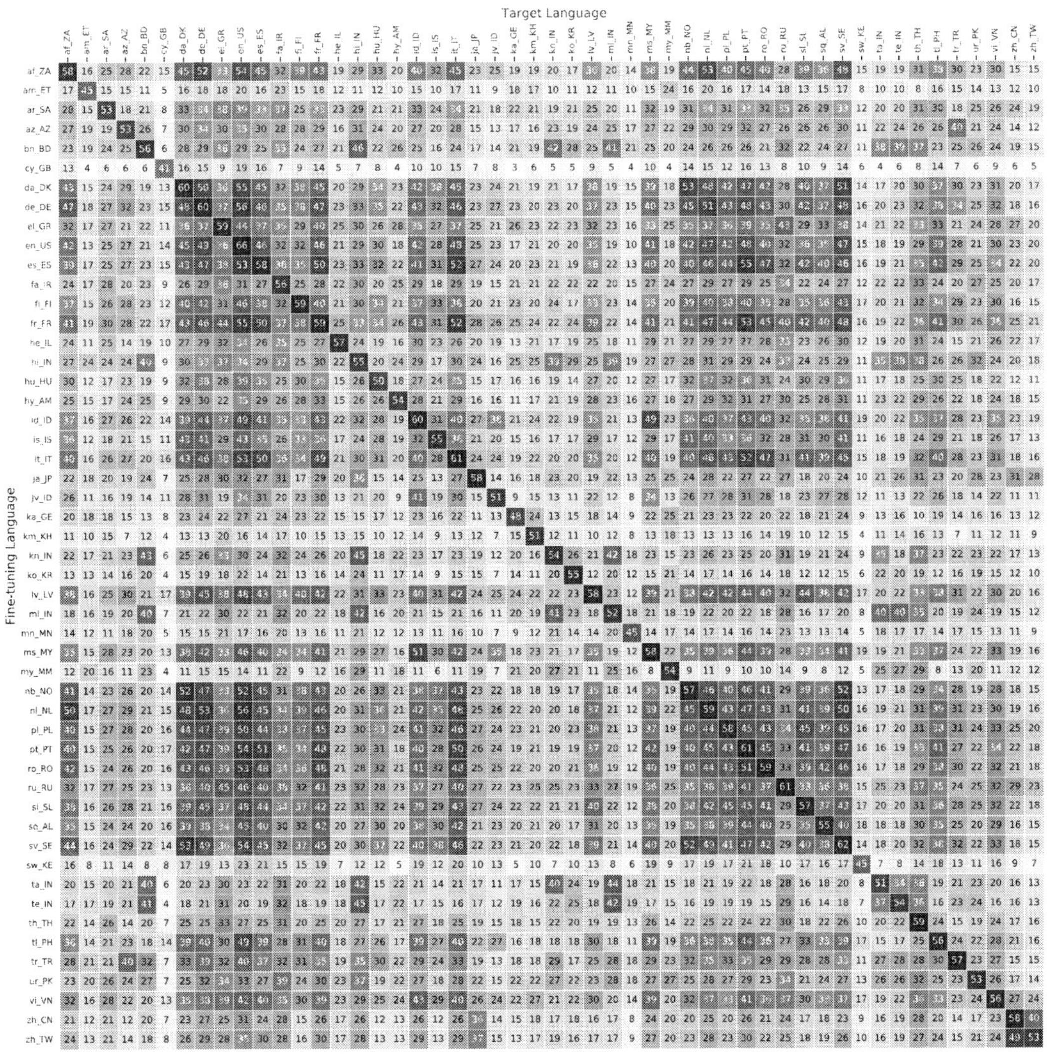

Figure 2: Zero-shot EM accuracies of individual ByT5-base models fine-tuned on a single language (y-axis) and evaluated on dev sets from all languages (x-axis).

(Schuster et al., 2019; Chen et al., 2020; Zenkel et al., 2020). More recent works removes this explicit alignment requirement (Dong and Lapata, 2018; Zhang et al., 2019; Wiseman et al., 2018). In our work, we use a label projection method based on pretrained language models (Nicosia et al., 2021) that reconstructs a full semantic parse from an utterance and a signature of the same parse.

6 Conclusions

In this paper, we evaluated ByT5 and mT5 (Xue et al., 2021) models in a massively multilingual semantic parsing task, showing that ByT5 is particularly competitive at smaller sizes. We have provided a map of the cross-lingual transfer for all the languages in MASSIVE and demonstrated that synthetic examples created with NMT are effective for building accurate semantic parsers.

Limitations

This work uses seq2seq models as parsers. Different output formats can yield better or worse results as shown in Paolini et al. (2021). We do not focus on tweaking formats or on modeling improvements such as constrained decoding for a more faithful generation. We adopt a compact output representation that reduces the text the model has to generate (and hallucinations) and gives us competitive results. In the cross-lingual transfer experiments, we train each model for a small fixed number of steps. If we train for longer, the representations start to change significantly and cross-lingual performances vary quite unpredictably. We leave for the future an investigation of the learning dynamics in this setting and the design of possible remedies.

References

Yoav Artzi and Luke Zettlemoyer. 2011. Bootstrapping semantic parsers from conversations. In *Proceedings of the 2011 Conference on Empirical Methods in Natural Language Processing*, pages 421–432.

Emanuele Bastianelli, Andrea Vanzo, Pawel Swietojanski, and Verena Rieser. 2020. SLURP: A spoken language understanding resource package. In *Proceedings of the 2020 Conference on Empirical Methods in Natural Language Processing (EMNLP)*, pages 7252–7262, Online. Association for Computational Linguistics.

Jonathan Berant, Andrew Chou, Roy Frostig, and Percy Liang. 2013. Semantic parsing on freebase from question-answer pairs. In *Proceedings of the 2013 conference on empirical methods in natural language processing*, pages 1533–1544.

Peter F. Brown, Stephen A. Della Pietra, Vincent J. Della Pietra, and Robert L. Mercer. 1993. The mathematics of statistical machine translation: Parameter estimation. *Computational Linguistics*, 19(2):263–311.

Yun Chen, Yang Liu, Guanhua Chen, Xin Jiang, and Qun Liu. 2020. Accurate word alignment induction from neural machine translation. In *Proceedings of the 2020 Conference on Empirical Methods in Natural Language Processing (EMNLP)*, pages 566–576, Online. Association for Computational Linguistics.

Jonathan H. Clark, Dan Garrette, Iulia Turc, and John Wieting. 2022. Canine: Pre-training an efficient tokenization-free encoder for language representation. *Transactions of the Association for Computational Linguistics*, 10:73–91.

Alexis Conneau, Kartikay Khandelwal, Naman Goyal, Vishrav Chaudhary, Guillaume Wenzek, Francisco Guzmán, Edouard Grave, Myle Ott, Luke Zettlemoyer, and Veselin Stoyanov. 2020. Unsupervised cross-lingual representation learning at scale. In *Proceedings of the 58th Annual Meeting of the Association for Computational Linguistics*, pages 8440–8451, Online. Association for Computational Linguistics.

Bosheng Ding, Junjie Hu, Lidong Bing, Mahani Aljunied, Shafiq Joty, Luo Si, and Chunyan Miao. 2022. GlobalWoZ: Globalizing MultiWoZ to develop multilingual task-oriented dialogue systems. In *Proceedings of the 60th Annual Meeting of the Association for Computational Linguistics (Volume 1: Long Papers)*, pages 1639–1657, Dublin, Ireland. Association for Computational Linguistics.

Li Dong and Mirella Lapata. 2018. Coarse-to-fine decoding for neural semantic parsing. In *Proceedings of the 56th Annual Meeting of the Association for Computational Linguistics (Volume 1: Long Papers)*, pages 731–742, Melbourne, Australia. Association for Computational Linguistics.

Kais Dukes. 2014. Semeval-2014 task 6: Supervised semantic parsing of robotic spatial commands. In *SemEval@ COLING*, pages 45–53.

Jack FitzGerald, Christopher Hench, Charith Peris, Scott Mackie, Kay Rottmann, Ana Sanchez, Aaron Nash, Liam Urbach, Vishesh Kakarala, Richa Singh, Swetha Ranganath, Laurie Crist, Misha Britan, Wouter Leeuwis, Gokhan Tur, and Prem Natarajan. 2022. Massive: A 1m-example multilingual natural language understanding dataset with 51 typologically-diverse languages.

Junjie Hu, Sebastian Ruder, Aditya Siddhant, Graham Neubig, Orhan Firat, and Melvin Johnson. 2020. XTREME: A massively multilingual multi-task benchmark for evaluating cross-lingual generalisation. In *Proceedings of the 37th International Conference on Machine Learning*, volume 119 of *Proceedings of Machine Learning Research*, pages 4411–4421. PMLR.

Faisal Ladhak, Esin Durmus, Claire Cardie, and Kathleen McKeown. 2020. WikiLingua: A new benchmark dataset for cross-lingual abstractive summarization. In *Findings of the Association for Computational Linguistics: EMNLP 2020*, pages 4034–4048, Online. Association for Computational Linguistics.

Guillaume Lample and Alexis Conneau. 2019. Cross-lingual language model pretraining. *CoRR*, abs/1901.07291.

Haoran Li, Abhinav Arora, Shuohui Chen, Anchit Gupta, Sonal Gupta, and Yashar Mehdad. 2021. MTOP: A comprehensive multilingual task-oriented semantic parsing benchmark. In *Proceedings of the 16th Conference of the European Chapter of the Association for Computational Linguistics: Main Volume*, pages 2950–2962, Online. Association for Computational Linguistics.

Zhaojiang Lin, Andrea Madotto, Genta Indra Winata, Peng Xu, Feijun Jiang, Yuxiang Hu, Chen Shi, and Pascale Fung. 2021. Bitod: A bilingual multi-domain dataset for task-oriented dialogue modeling. In *Thirty-fifth Conference on Neural Information Processing Systems Datasets and Benchmarks Track (Round 1)*.

Olga Majewska, Evgeniia Razumovskaia, Edoardo Maria Ponti, Ivan Vulić, and Anna Korhonen. 2022. Cross-lingual dialogue dataset creation via outline-based generation.

Mehrad Moradshahi, Giovanni Campagna, Sina Semnani, Silei Xu, and Monica Lam. 2020. Localizing open-ontology QA semantic parsers in a day using machine translation. In *Proceedings of the 2020 Conference on Empirical Methods in Natural Language Processing (EMNLP)*, pages 5970–5983, Online. Association for Computational Linguistics.

Massimo Nicosia, Zhongdi Qu, and Yasemin Altun. 2021. Translate & Fill: Improving zero-shot multilingual semantic parsing with synthetic data. In *Findings of the Association for Computational Linguistics: EMNLP 2021*, pages 3272–3284, Punta Cana, Dominican Republic. Association for Computational Linguistics.

Franz Josef Och and Hermann Ney. 2000. Improved statistical alignment models. In *Proceedings of the 38th Annual Meeting of the Association for Computational Linguistics*, pages 440–447, Hong Kong. Association for Computational Linguistics.

Franz Josef Och and Hermann Ney. 2003. A systematic comparison of various statistical alignment models. *Computational Linguistics*, 29(1):19–51.

Giovanni Paolini, Ben Athiwaratkun, Jason Krone, Jie Ma, Alessandro Achille, RISHITA ANUBHAI, Cicero Nogueira dos Santos, Bing Xiang, and Stefano Soatto. 2021. Structured prediction as translation between augmented natural languages. In *International Conference on Learning Representations*.

Colin Raffel, Noam Shazeer, Adam Roberts, Katherine Lee, Sharan Narang, Michael Matena, Yanqi Zhou, Wei Li, and Peter J. Liu. 2020. Exploring the limits of transfer learning with a unified text-to-text transformer. *Journal of Machine Learning Research*, 21(140):1–67.

Sebastian Schuster, Sonal Gupta, Rushin Shah, and Mike Lewis. 2019. Cross-lingual transfer learning for multilingual task oriented dialog. In *Proceedings of the 2019 Conference of the North American Chapter of the Association for Computational Linguistics: Human Language Technologies, Volume 1 (Long and Short Papers)*, pages 3795–3805. Association for Computational Linguistics.

Tom Sherborne, Yumo Xu, and Mirella Lapata. 2020. Bootstrapping a crosslingual semantic parser. In *Findings of the Association for Computational Linguistics: EMNLP 2020*, pages 499–517, Online. Association for Computational Linguistics.

Yi Tay, Vinh Q. Tran, Sebastian Ruder, Jai Gupta, Hyung Won Chung, Dara Bahri, Zhen Qin, Simon Baumgartner, Cong Yu, and Donald Metzler. 2022. Charformer: Fast character transformers via gradient-based subword tokenization. In *International Conference on Learning Representations*.

Iulia Turc, Kenton Lee, Jacob Eisenstein, Ming-Wei Chang, and Kristina Toutanova. 2021. Revisiting the primacy of english in zero-shot cross-lingual transfer. *ArXiv*, abs/2106.16171.

Stephan Vogel, Hermann Ney, and Christoph Tillmann. 1996. HMM-based word alignment in statistical translation. In *COLING 1996 Volume 2: The 16th International Conference on Computational Linguistics*.

Xinyi Wang, Sebastian Ruder, and Graham Neubig. 2021. Multi-view subword regularization. In *Proceedings of the 2021 Conference of the North American Chapter of the Association for Computational Linguistics: Human Language Technologies*, pages 473–482, Online. Association for Computational Linguistics.

Sam Wiseman, Stuart Shieber, and Alexander Rush. 2018. Learning neural templates for text generation. In *Proceedings of the 2018 Conference on Empirical Methods in Natural Language Processing*, pages 3174–3187, Brussels, Belgium. Association for Computational Linguistics.

Weijia Xu, Batool Haider, and Saab Mansour. 2020. End-to-end slot alignment and recognition for cross-lingual NLU. In *Proceedings of the 2020 Conference on Empirical Methods in Natural Language Processing (EMNLP)*, pages 5052–5063, Online. Association for Computational Linguistics.

Linting Xue, Noah Constant, Adam Roberts, Mihir Kale, Rami Al-Rfou, Aditya Siddhant, Aditya Barua, and Colin Raffel. 2021. mt5: A massively multilingual pre-trained text-to-text transformer.

Thomas Zenkel, Joern Wuebker, and John DeNero. 2020. End-to-end neural word alignment outperforms GIZA++. In *Proceedings of the 58th Annual Meeting of the Association for Computational Linguistics*, pages 1605–1617, Online. Association for Computational Linguistics.

Xiang Zhang, Shizhu He, Kang Liu, and Jun Zhao. 2019. AdaNSP: Uncertainty-driven adaptive decoding in neural semantic parsing. In *Proceedings of the 57th Annual Meeting of the Association for Computational Linguistics*, pages 4265–4270, Florence, Italy. Association for Computational Linguistics.

A Comparing NMT with Gold Translations

In Table 7, we compare how many times the NMT translated utterances match the gold translations produced by professional translators. We restrict the match to utterances that have been translated and not localized in the target language, since NMT cannot perform the localization step. In addition, we preprocess all compared utterances with unicode normalization, we strip whitespaces and punctuation. In general, indic locales have higher match rates compared to other locales. Please also note that we translate English to pt_BR (Brazilian Portuguese) and this explains the low match for pt_PT.

B Intent Accuracy Performance

In Table 8, we report the accuracy for each individual intent on the union of the test set examples from all languages using ByT5-xxl + TAF.

C Performance on all Languages

In Table 9, we report Exact Match on all the 51 languages, for the three different experimental setups described in Section 3, across two models (mT5 and ByT5) and two model sizes (base and xxl).

Language	NMT vs Gold Translations Matches (%)	(#)	Non-localized sentences (#)
kn_IN	68.7	6524	9497
te_IN	54.1	4841	8941
bn_BD	52.6	4458	8471
ta_IN	48.3	4301	8898
hi_IN	46.5	4101	8827
nl_NL	38.5	3878	10 070
fr_FR	36.0	3736	10 385
ml_IN	34.7	2985	8607
tl_PH	34.0	3397	10 000
af_ZA	32.8	3160	9640
tr_TR	32.1	2998	9330
sw_KE	26.1	2336	8965
sv_SE	25.9	2465	9504
nb_NO	23.8	2402	10 083
vi_VN	21.6	2000	9255
ms_MY	21.6	1880	8702
jv_ID	21.1	1947	9208
pl_PL	21.0	2017	9618
da_DK	20.4	1933	9470
id_ID	20.4	1882	9227
es_ES	19.5	1876	9596
zh_CN	19.0	1661	8727
zh_TW	18.2	1638	8976
it_IT	17.9	1596	8916
fi_FI	17.5	1669	9558
ru_RU	17.4	1550	8912
hy_AM	16.9	1809	10 707
is_IS	16.1	1491	9270
km_KH	16.1	1491	9276
cy_GB	15.9	1578	9936
sl_SL	14.7	1313	8913
am_ET	14.6	1267	8658
hu_HU	14.5	1331	9198
ur_PK	14.4	1260	8761
de_DE	14.2	1422	9992
lv_LV	12.4	1071	8650
he_IL	12.3	1123	9159
sq_AL	12.2	1035	8460
az_AZ	12.1	1102	9081
th_TH	11.7	1041	8894
ro_RO	10.9	1001	9197
el_GR	10.5	934	8879
pt_PT	9.9	934	9392
ar_SA	9.9	871	8814
mn_MN	8.9	785	8826
fa_IR	8.3	718	8686
ja_JP	7.4	704	9487
ka_GE	7.4	701	9528
ko_KR	3.9	341	8804
my_MM	2.0	171	8765

Table 7: Number of verbatim matches between Gold translation and NMT translations.

Intent	IA	Support
GENERAL_GREET	19.6	51
MUSIC_SETTINGS	27.1	306
AUDIO_VOLUME_OTHER	54.9	306
GENERAL_QUIRKY	55.6	8619
IOT_HUE_LIGHTON	61.4	153
MUSIC_DISLIKENESS	74.5	204
DATETIME_CONVERT	75.6	765
IOT_WEMO_ON	76.3	510
PLAY_AUDIOBOOK	78.0	2091
TRANSPORT_QUERY	78.1	2601
RECOMMENDATION_EVENTS	78.3	2193
RECOMMENDATION_MOVIES	79.2	1020
CALENDAR_QUERY	80.6	6426
QA_FACTOID	82.4	7191
IOT_HUE_LIGHTUP	82.5	1377
LISTS_QUERY	82.6	2601
AUDIO_VOLUME_UP	83.0	663
SOCIAL_QUERY	83.9	1275
MUSIC_QUERY	84.0	1785
EMAIL_ADDCONTACT	84.5	612
MUSIC_LIKENESS	84.7	1836
EMAIL_QUERYCONTACT	84.8	1326
TAKEAWAY_QUERY	85.0	1785
LISTS_CREATEORADD	85.6	1989
QA_DEFINITION	86.3	2907
LISTS_REMOVE	86.3	2652
COOKING_RECIPE	86.6	3672
NEWS_QUERY	86.9	6324
PLAY_MUSIC	87.1	8976
TAKEAWAY_ORDER	87.3	1122
IOT_HUE_LIGHTDIM	87.4	1071
PLAY_PODCASTS	87.6	3213
PLAY_GAME	87.7	1785
ALARM_SET	89.5	2091
PLAY_RADIO	90.0	3672
CALENDAR_SET	90.2	10 659
RECOMMENDATION_LOCATIONS	90.4	1581
QA_MATHS	90.7	1275
AUDIO_VOLUME_DOWN	90.7	561
SOCIAL_POST	91.1	4131
IOT_WEMO_OFF	91.3	918
AUDIO_VOLUME_MUTE	91.7	1632
ALARM_QUERY	91.8	1734
GENERAL_JOKE	92.0	969
EMAIL_QUERY	93.0	6069
TRANSPORT_TICKET	93.1	1785
CALENDAR_REMOVE	93.4	3417
EMAIL_SENDEMAIL	94.0	5814
IOT_CLEANING	94.2	1326
WEATHER_QUERY	94.6	7956
IOT_HUE_LIGHTOFF	94.8	2193
TRANSPORT_TAXI	95.3	1173
IOT_HUE_LIGHTCHANGE	95.4	1836
ALARM_REMOVE	95.5	1071
QA_STOCK	95.6	1326
DATETIME_QUERY	95.8	4488
TRANSPORT_TRAFFIC	96.3	765
QA_CURRENCY	96.6	1989
IOT_COFFEE	97.9	1836

Table 8: IA of the ByT5-xxl+TAF model for all intents (all languages).

	Zero Shot				Synthetic (TAF)				Gold			
	base		xxl		base		xxl		base		xxl	
Language	mT5	ByT5	mT5	ByT5	mT5	ByT5	mT5	ByT5	mT5	ByT5	mT5	ByT5
af_ZA	21.6	**51.1**	58.0	**59.7**	53.7	**64.7**	65.6	**66.8**	59.4	**68.5**	65.9	**69.3**
am_ET	4.7	**15.9**	**40.7**	22.0	40.8	**54.4**	**61.2**	61.0	48.7	**61.3**	62.0	**65.8**
ar_SA	14.6	**27.8**	**43.6**	23.3	45.9	**56.1**	60.1	**60.5**	52.3	**64.7**	61.1	**66.0**
az_AZ	8.9	**31.2**	**41.8**	34.0	46.4	**61.6**	61.9	**63.6**	57.0	**69.0**	62.6	**69.6**
bn_BD	10.8	**19.5**	**45.9**	25.3	51.0	**62.1**	64.3	**65.6**	57.6	**67.6**	64.6	**69.5**
cy_GB	5.9	**16.4**	**42.8**	40.2	35.7	**56.1**	61.5	**64.2**	42.1	**65.3**	61.4	**69.2**
da_DK	30.2	**53.1**	**60.9**	54.2	57.8	**67.5**	67.3	**68.7**	64.4	**71.7**	67.9	**71.3**
de_DE	28.3	**55.3**	**59.8**	59.5	60.2	**67.8**	67.5	**68.8**	64.1	**70.4**	68.0	**70.2**
el_GR	17.4	**31.5**	**57.2**	27.9	55.5	**64.2**	65.5	**66.6**	62.0	**68.3**	66.6	**68.7**
en_US	65.5	**72.2**	**74.0**	73.3	68.5	**72.6**	**73.7**	73.0	68.9	**72.7**	**73.3**	72.6
es_ES	26.1	**50.8**	**55.6**	52.2	58.7	**65.1**	65.0	**65.9**	61.1	**67.2**	65.9	**66.2**
fa_IR	17.6	**32.8**	**54.4**	24.0	54.9	**62.2**	63.2	**64.4**	59.9	**69.1**	63.4	**69.7**
fi_FI	16.3	**36.9**	**52.5**	47.4	51.2	**65.9**	65.6	**68.2**	59.4	**71.1**	66.8	**71.5**
fr_FR	29.9	**53.5**	**58.5**	54.3	59.3	**64.4**	65.1	**65.6**	62.3	**66.5**	65.8	**67.2**
he_IL	9.7	**21.0**	**40.4**	24.0	50.1	**59.4**	61.0	**63.2**	57.5	**67.3**	62.3	**68.4**
hi_IN	14.1	**26.3**	**52.9**	26.2	54.4	**62.6**	64.2	**64.4**	59.3	**66.5**	64.5	**67.2**
hu_HU	17.5	**33.5**	**45.3**	32.9	51.8	**62.2**	64.2	64.2	58.2	**68.5**	65.2	**69.5**
hy_AM	11.7	**20.5**	**44.6**	24.7	49.8	**58.4**	60.3	**62.2**	57.8	**67.7**	61.7	**68.9**
id_ID	24.1	**48.3**	58.6	**61.5**	59.0	**64.6**	65.5	**67.1**	63.4	**68.8**	66.2	**69.0**
is_IS	11.6	**32.1**	**47.2**	31.7	47.6	**60.9**	63.4	**65.9**	54.6	**68.5**	63.4	**69.6**
it_IT	25.3	**52.5**	59.5	59.5	57.2	**63.0**	64.6	**65.5**	60.2	**67.6**	65.7	**67.3**
ja_JP	**26.8**	23.3	**46.6**	29.3	51.0	**55.6**	57.3	**58.8**	60.5	**65.8**	58.7	**67.0**
jv_ID	10.7	**22.9**	45.8	**46.2**	42.5	**58.9**	62.1	**63.9**	48.5	**66.5**	62.6	**68.5**
ka_GE	9.7	**17.9**	**39.9**	22.1	45.4	**52.9**	54.8	**57.1**	54.5	**63.8**	56.2	**66.8**
km_KH	11.4	**18.0**	**44.8**	23.6	39.2	**51.8**	51.7	**55.7**	54.7	**63.8**	54.3	**67.0**
kn_IN	8.8	**20.2**	**41.9**	25.4	47.4	**58.6**	55.8	**61.7**	52.1	**63.8**	56.6	**65.8**
ko_KR	11.0	**16.3**	**49.8**	24.8	54.1	**61.5**	65.6	**65.8**	60.2	**68.7**	66.4	**70.3**
lv_LV	11.6	**40.3**	**51.9**	33.7	52.4	**61.2**	63.0	**64.6**	59.0	**69.6**	64.1	**70.4**
ml_IN	10.1	**19.4**	**41.2**	25.8	47.9	**55.3**	55.0	**58.5**	59.4	**68.2**	55.6	**69.2**
mn_MN	7.4	**13.4**	**38.9**	22.2	46.9	**57.0**	60.2	**62.7**	53.8	**66.1**	61.5	**68.7**
ms_MY	21.7	**45.0**	54.8	**59.9**	57.1	**65.7**	67.7	**68.0**	60.6	**69.3**	68.4	**68.9**
my_MM	10.7	**13.8**	**48.7**	23.1	51.5	**59.8**	61.9	**66.1**	59.3	**68.8**	64.3	**72.6**
nb_NO	26.9	**50.6**	**60.7**	56.3	60.7	**68.0**	68.8	**70.2**	65.0	**70.5**	69.9	**70.7**
nl_NL	28.3	**55.2**	60.1	**63.3**	60.2	**66.5**	67.4	**67.5**	64.7	**68.4**	68.3	**70.0**
pl_PL	19.0	**47.1**	**50.7**	46.0	56.2	**61.8**	62.0	**63.3**	59.7	**65.9**	62.5	**66.5**
pt_PT	28.1	**52.0**	**60.8**	50.6	61.5	**65.9**	66.8	**67.6**	63.6	**68.7**	67.5	**68.2**
ro_RO	22.8	**45.7**	**57.4**	52.7	55.8	**64.5**	65.7	**67.1**	60.2	**68.5**	65.9	**69.6**
ru_RU	19.0	**26.1**	**49.0**	26.1	56.9	**61.6**	63.5	**63.8**	63.5	**68.8**	64.0	**69.5**
sl_SL	15.8	**43.7**	**52.8**	47.8	53.2	**63.5**	64.5	**64.8**	57.7	**68.0**	64.5	**68.8**
sq_AL	15.3	**42.1**	**48.0**	39.9	48.8	**61.1**	61.2	**63.5**	54.2	**68.9**	61.3	**68.5**
sv_SE	26.0	**54.4**	**61.8**	53.0	62.6	**70.1**	70.6	**71.1**	65.9	**72.0**	71.2	**71.5**
sw_KE	9.6	**15.6**	**44.0**	41.9	44.2	**58.7**	58.2	**59.6**	48.0	**66.3**	58.6	**66.8**
ta_IN	10.9	**19.9**	**41.1**	24.3	48.2	**55.5**	56.4	**58.3**	56.6	**64.9**	58.0	**66.0**
te_IN	7.8	**21.6**	**46.4**	25.1	43.6	**60.0**	55.4	**62.7**	51.4	**65.0**	55.1	**67.5**
th_TH	21.8	**31.3**	**55.0**	26.8	47.4	**62.1**	62.2	**66.9**	63.2	**72.0**	64.6	**74.2**
tl_PH	18.9	**42.0**	56.9	**58.7**	53.2	**62.4**	65.7	**66.1**	56.7	**66.5**	66.5	**68.5**
tr_TR	14.4	**35.2**	**48.4**	38.5	51.6	**64.9**	65.5	**66.2**	58.5	**69.4**	65.5	**69.4**
ur_PK	9.7	**22.7**	**49.2**	22.8	50.5	**59.5**	61.5	**61.9**	54.1	**63.3**	62.6	**65.7**
vi_VN	15.1	**35.1**	**55.9**	36.4	49.8	**57.5**	61.0	**62.3**	55.5	**67.0**	62.1	**68.2**
zh_CN	**22.1**	17.3	**31.7**	24.1	45.6	**54.1**	53.0	**57.9**	60.8	**65.9**	54.9	**66.6**
zh_TW	**21.2**	16.5	**32.4**	24.2	45.2	**51.8**	52.0	**54.5**	58.2	**62.2**	53.8	**63.9**
Average	17.7	**33.5**	**50.2**	38.3	51.8	**61.2**	62.5	**64.2**	58.2	**67.5**	63.3	**68.7**

Table 9: *T5 parsers Exact Match on individual languages in the Zero-Shot, TAF and Gold settings.

HIT-SCIR at MMNLU-22: Consistency Regularization for Multilingual Spoken Language Understanding

Bo Zheng, Zhouyang Li, Fuxuan Wei, Qiguang Chen, Libo Qin, Wanxiang Che*
Harbin Institute of Technology
{bzheng,zhouyangli,fxwei,qgchen,lbqin,car}@ir.hit.edu.cn

Abstract

Multilingual spoken language understanding (SLU) consists of two sub-tasks, namely intent detection and slot filling. To improve the performance of these two sub-tasks, we propose to use consistency regularization based on a hybrid data augmentation strategy. The consistency regularization enforces the predicted distributions for an example and its semantically equivalent augmentation to be consistent. We conduct experiments on the MASSIVE dataset under both full-dataset and zero-shot settings. Experimental results demonstrate that our proposed method improves the performance on both intent detection and slot filling tasks. Our system[1] ranked 1st in the MMNLU-22 competition under the full-dataset setting.

1 Introduction

The MMNLU-22 evaluation focuses on the problem of multilingual natural language understanding. It is based on the MASSIVE dataset (FitzGerald et al., 2022), a multilingual spoken language understanding (SLU) dataset with two sub-tasks, including *intent detection* and *slot filling*. Specifically, given a virtual assistant utterance in an arbitrary language, the model is designed to predict the corresponding intent label and extract the slot results. An English example is illustrated in Figure 1.

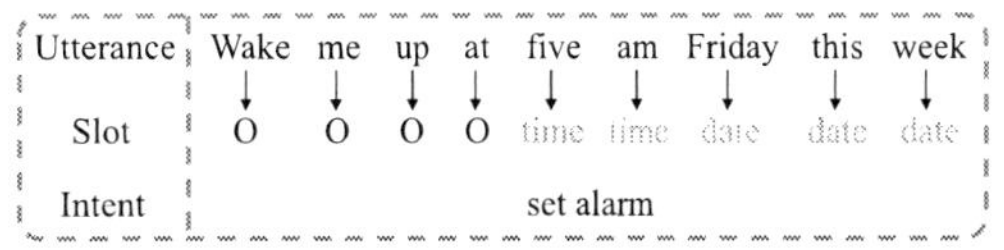

Figure 1: An English example from the MASSIVE dataset. The slot label 'O' stands for the 'Other' label.

Fine-tuning pre-trained cross-lingual language models allows task-specific supervision to be shared and transferred across languages (Conneau and Lample, 2019; Conneau et al., 2020; Xue et al., 2021). This motivates the two setting for the MMNLU-22 evaluation, namely the *full-dataset* setting and the *zero-shot* setting. Participants are allowed to use training data in all languages under the full-dataset setting, while they can only access the English training data under the zero-shot setting. The latter is also called zero-shot cross-lingual SLU in previous work (Qin et al., 2020, 2022).

Cross-lingual data augmentation methods have been proven effective to improve cross-lingual transferability, e.g., code-switch substitution (Qin et al., 2020) and machine translation (Conneau and Lample, 2019; Singh et al., 2019). Most previous work directly utilizes the data augmentations as additional training data for fine-tuning. However, they ignore the inherent correlation between the original example and its semantically equivalent augmentation, which can be fully exploited with the *consistency regularization* (Zheng et al., 2021b). The consistency regularization enforces the model predictions to be more consistent for semantic-preserving augmentations.

Motivated by this, we propose to apply consistency regularization based on a hybrid data augmentation strategy, including data augmentation of machine translation and subword sampling (Kudo, 2018). We use machine translation augmentation to align the model predictions of the intent detection task. Meanwhile, subword sampling augmentation is used to align the model predictions of both intent detection and slot filling tasks. The proposed method consistently improves the SLU performance on the MASSIVE dataset under both full-dataset and zero-shot settings. It is worth mentioning that our system ranked 1st in the MMNLU-22 competition under the full-dataset setting. We achieved an exact match accuracy of 49.65 points, outperforming the 2nd system by 1.02 points.

*Email corresponding.
[1]The code will be available at https://github.com/bozheng-hit/MMNLU-22-HIT-SCIR.

Proceedings of the Massively Multilingual Natural Language Understanding Workshop (MMNLU-22), pages 35 - 41
December 7, 2022

2 Background

2.1 Task Description

The task of SLU is that given an utterance with a word sequence $\boldsymbol{x} = (x_1, ..., x_n)$ with length n. The model is required to solve two sub-tasks. The intent detection task can be seen as an utterance classification task to decide the intent label o^I, and the slot filling task is a sequence labeling task that generates a slot label for each word in the utterance to obtain the slot sequence $\boldsymbol{o}^S = (o_1^S, ..., o_n^S)$.

2.2 Dataset Description

The MASSIVE dataset is composed of realistic, human-created virtual assistant utterance text spanning 51 languages, 60 intents, 55 slot types, and 18 domains (FitzGerald et al., 2022). There are 11,514 training utterances for each language. For the full-dataset setting, all training data can be used. For the zero-shot setting, only English training data can be used, yet we can translate them into other languages using commercial translators. There are 2,033, 2,974, and 3,000 utterances for each language in the development, test, and evaluation set, respectively. The average performance in all languages should be reported under the full-dataset setting. Meanwhile, the average performance in all languages except English should be reported under the zero-shot setting.

2.3 Related Work

Pre-trained cross-lingual language models (Conneau and Lample, 2019; Conneau et al., 2020; Chi et al., 2021a,b, 2022; Xue et al., 2021) encode different languages into universal representations and significantly improve cross-lingual transferability. These models usually consist of a multilingual vocabulary (Conneau and Lample, 2019; Conneau et al., 2020; Xue et al., 2021; Zheng et al., 2021a) and a Transformer model (Vaswani et al., 2017).

A simple yet effective way to improve cross-lingual fine-tuning is to populate the training data with cross-lingual data augmentation (Conneau et al., 2020). Singh et al. (2019) replace a segment of source language input text with its translation in another language as data augmentation. Qin et al. (2020) randomly replace words in the source-language training example with target-language words using the bilingual dictionaries. Then the model is fine-tuned on the generated code-switched data. Instead of directly treating cross-lingual data augmentation as extra training data, Zheng et al. (2021b) proposed to better use data augmentations based on consistency regularization.

3 Method

Given the input utterance $\boldsymbol{x} = (x_1, ..., x_n)$ with length n and the corresponding intent label o^I and slot labels $\boldsymbol{o}^S = (o_1^S, ..., ...o_n^S)$ from training corpus $\mathcal{D}$, we define the loss for the two sub-tasks of SLU in our fine-tuning process as:

$$\mathcal{L}_I = \sum_{(\boldsymbol{x}, o^I) \in \mathcal{D}} \text{CE}(f_I(\boldsymbol{x}), o^I),$$

$$\mathcal{L}_S = \sum_{(\boldsymbol{x}, \boldsymbol{o}^S) \in \mathcal{D}} \text{CE}(f_S(\boldsymbol{x}), \boldsymbol{o}^S),$$

where $\mathcal{L}_I$ and $\mathcal{L}_S$ stand for the intent detection task and the slot filling task, $f_I(\cdot)$ and $f_S(\cdot)$ denote the model which predicts task-specific probability distributions for the input example x, $\text{CE}(\cdot, \cdot)$ denotes cross-entropy loss.

3.1 Consistency Regularization

In order to make better use of data augmentations, we introduce the consistency regularization used in Zheng et al. (2021b), which encourages consistent predictions for an example and its semantically equivalent augmentation. We apply consistency regularization on intent detection and slot filling tasks, which is defined as follows:

$$\mathcal{R}_I = \sum_{\boldsymbol{x} \in \mathcal{D}} \text{KL}(f_I(\boldsymbol{x}) \| f_I(\mathcal{A}(\boldsymbol{x}, z))),$$

$$\mathcal{R}_S = \sum_{\boldsymbol{x} \in \mathcal{D}} \text{KL}(f_S(\boldsymbol{x}) \| f_S(\mathcal{A}(\boldsymbol{x}, z))),$$

$$\text{KL}_\text{S}(P \| Q) = \text{KL}(\text{stopgrad}(P) \| Q) + \text{KL}(\text{stopgrad}(Q) \| P)$$

where $\text{KL}_\text{S}(\cdot \| \cdot)$ is the symmertrical Kullback-Leibler divergence, $\mathcal{A}(\boldsymbol{x}, z)$ denotes the augmented version of input utterance x with data augmentation strategy z. The regularizer encourages the predicted distributions of the original training example and its augmented version to agree with each other. The $\text{stopgrad}(\cdot)$ operation[2] is used to stop back-propagating gradients, which is also employed in (Jiang et al., 2020; Liu et al., 2020; Zheng et al., 2021b).

3.2 Data Augmentations

We consider two types of data augmentation strategies for our consistency regularization method, including subword sampling and machine translation.

[2]Implemented by `.detach()` in PyTorch.

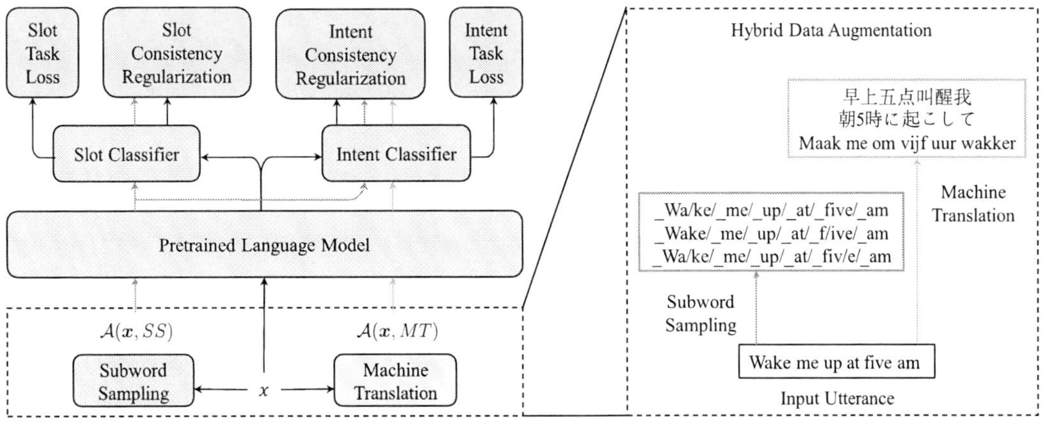

Figure 2: Illustration of our fine-tuning framework. 'MT' denotes machine translation augmentation and 'SS' denotes subword sampling augmentation.

3.2.1 Subword Sampling

Subword sampling is to generate multiple subword sequences from the original text as data augmentation. We apply the on-the-fly subword sampling algorithm from the unigram language model (Kudo, 2018) in SentencePiece (Kudo and Richardson, 2018). The output distributions of slot labels are generated on the first subword of each word in the input utterance. Therefore, the subword sampling augmentation can be used to align the output distribution of both intent detection and slot filling tasks.

3.2.2 Machine Translation

Machine translation is a common and effective data augmentation strategy in the cross-lingual scenario (Conneau and Lample, 2019; Singh et al., 2019). Due to the difficulty of accessing ground-truth labels in translation examples, machine translation can not be an available data augmentation strategy in the slot filling task. To improve the quality of our translations, we employ a variety of approaches (See Section 4.2). Unlike subword sampling, the output distributions of slot labels between the translation pairs can not be aligned. Thus, we only use machine translation to align the output distributions of the intent detection task.

3.3 Consistency Regularization based on Hybrid Data Augmentations

We illustrate our fine-tuning framework in Figure 2. We propose to use consistency regularization based on a hybrid data augmentation strategy, which includes data augmentation of machine translation and subword sampling. During the training process, we perform task fine-tuning and consistency regularization for an input example simultaneously. Then the final training loss is defined as follows:

$$\mathcal{L} = \mathcal{L}_I + \lambda_1 \mathcal{L}_S + \lambda_2 \mathcal{R}_I + \lambda_3 \mathcal{R}_S$$

where λ_1 is the slot loss coefficient, λ_2 and λ_3 are the corresponding weights of the consistency regularization for two tasks. We sample different data augmentation for the input example with the pre-defined distribution.

4 Experiments

4.1 Experimental Setup

We consider two types of pre-trained cross-lingual language models, which are encoder-only models and Text-to-Text models.

We use XLM-Align Base (Chi et al., 2021b) for the encoder-only model setting. We use a two-layer feed-forward network with a 3,072 hidden size. We use the first representation of sentences "<s>" for the intent detection task and the first subword of each word for the slot filling task.

We use mT5 Base (Xue et al., 2021) for the Text-to-Text model setting. We follow FitzGerald et al. (2022) to concatenate "Annotate: " and the unlabeled input utterance as the input of the encoder, and generate the text concatenation of the intent label and the slot labels as the decoder output. The labels are separated with white spaces and then tokenized into subwords.

We select the model that performs the best on the development dataset to run prediction on the test and evaluation dataset. We mainly select the batch size in $[32, 64, 128, 256]$, dropout rate in

Text Type	Text Content	Slot Translation	Text Translation	Aligned or Not
Plain Text	Wake me up at five am Friday this week	five am: 凌晨五点	本周周五凌晨五点叫我起床	Yes
Text with Slots in Brackets	Wake me up at [five am] [Friday this week]	Friday this week: 本周周五	在[凌晨五点][本周星期五]叫醒我	No
Plain Text	set an alarm for two hours from now	two hours from now:	从现在开始设置两个小时的闹钟	No
Text with Slots in Brackets	set an alarm for [two hours from now]	从现在起两小时后	设置[从现在起两小时后]的闹钟	Yes

Table 1: Examples of aligning slots into machine translations.

Model	Test Set			Evaluation Set		
	Intent Acc	Slot F1	EMA	Intent Acc	Slot F1	EMA
XLM-R Base	85.10	73.60	63.69	-	-	-
XLM-Align Base	**86.16**	76.36	66.42	-	-	-
mT5 Base Text-to-Text	85.33	**76.77**	**66.64**	-	-	-
XLM-Align Base + Ours	87.12	77.99	68.76	85.00	68.45	48.64
mT5 Base Text-to-Text + Ours	**87.60**	**78.22**	**69.60**	**85.10**	**69.08**	**49.65**

Table 2: Test and evaluation results on the MASSIVE dataset under the full-dataset setting. Results of XLM-R Base and mT5 Base Text-to-Text are taken from FitzGerald et al. (2022).

$[0.05, 0.1, 0.15]$, and the hyper-parameters used in our proposed method, including slot loss coefficient λ_1 in $[1, 2, 4]$, weights of consistency regularization λ_2 and λ_3 in $[2, 3, 5, 10]$. We select the learning rate in $[5e^{-5}, 8e^{-5}, 1e^{-4}]$ for Text-to-Text models. As for encoder-only models, we select the learning rate in $[4e^{-6}, 6e^{-6}, 8e^{-6}]$.

4.2 Data Processing

For the full-dataset setting, we use examples with the same id in different languages as machine translation augmentation in our fine-tuning framework. For the zero-shot setting, we translated the entire English training set into 50 languages using commercial translation APIs, such as DeepL translator and Google translator. These translations refer to plain text translations and can be used for intent detection training and consistency regularization.

We used two methods to obtain a translated example that aligned at the slot level. One is based on the plain text translation. Each slot value in an English training example is translated into a target language. If the translation results of each slot can be found in the plain text translation, a slot-aligned translation is obtained. The other is based on the annotated English training examples. We translated the annotated English training example with brackets for slot values (without slot type in brackets). Using brackets explicitly allows the translator to align slots to consecutive spans. And we also translated each slot value into the target language. If the translation result of each slot can be found in the annotated utterance translation, we obtain a slot alignment example after removing the brackets.

In practice, slot-aligned examples based on plain text translations are preferred as the final result of the slot alignment. If no such example is available, we use the slot-aligned results from annotated translations. Examples of slot alignment are shown in Table 1. For those plain text translations where we can not align the slot labels, we only use them for the training of the intent detection task.

4.3 Evaluation Metrics

The evaluation in competition is mainly conducted using three metrics:

- Exact Match Accuracy (EMA): The percentage of utterance-level predictions where the intent and all slots are exactly correct.
- Intent Accuracy (Intent Acc): The percentage of predictions in which the intent is correct.
- Slot Micro F1 (Slot F1): The micro-averaged F1 score is calculated over all slots.

4.4 Results

Table 2 shows our results on the MASSIVE dataset under the full-set setting. We tried different cross-lingual pre-trained language models under the baseline setting. Among them, XLM-Align Base performs the best on the intent detection task, while the mT5 Base Text-to-Text model performs the best on the slot filling task and exact match accuracy. When applying our consistency regularization method, the mT5 Base Text-to-Text model outperforms the XLM-Align Base model by 0.84 points and 0.99 points on exact match accuracy on the test dataset and the evaluation set, respectively. Meanwhile, compared to the baseline model, using consistency regularization achieves an absolute

Model	Test Set			Evaluation Set		
	Intent Acc	Slot F1	EMA	Intent Acc	Slot F1	EMA
XLM-R Base	70.62	50.27	38.70	-	-	-
XLM-Align Base	**68.49**	**54.69**	**40.91**	-	-	-
mT5 Base Text-to-Text	62.92	44.77	34.72	-	-	-
XLM-Align Base + Ours	85.12	71.27	62.18	83.18	62.84	43.05
XLM-Align Base + Ours + KD	**85.76**	**73.55**	**64.44**	**83.89**	**64.60**	**44.84**
mT5 Base Text-to-Text + Ours	84.58	69.24	60.59	82.56	60.00	40.93

Table 3: Test and evaluation results on the MASSIVE dataset under the zero-shot setting. Results of XLM-R Base and mT5 Base Text-to-Text are taken from FitzGerald et al. (2022).

Model	Intent Acc	Slot F1	EMA
XLM-Align Base + Ours	87.12	**77.99**	**68.76**
- Subword Sampling	**87.50**	76.08	67.40
- Consistency Regularization	86.16	76.32	66.57

Table 4: Ablation studies on the MASSIVE test dataset under the full-dataset setting.

Model	Intent Acc	Slot F1	EMA
XLM-Align Base + Ours	85.12	**71.27**	**62.18**
- Subword Sampling	**85.14**	69.52	60.94
- Machine Translation	72.27	58.37	45.50
- Consistency Regularization	83.90	69.37	59.95

Table 5: Ablation studies on the MASSIVE test dataset under the zero-shot setting.

2.96-point improvement on exact match accuracy with the mT5 Base Text-to-Text model.

Table 3 shows our results on the MASSIVE dataset under the zero-shot setting. For the baseline models, XLM-Align Base performs the best on all three metrics. Difference from the full-dataset setting, mT5 Base Text-to-Text models perform poorly under the zero-shot setting. We attribute it to the fact that Text-to-Text models strongly rely on the training data quality since most of the training data under the zero-shot setting are obtained with machine translation systems. When applying our consistency regularization method, the XLM-Align Base model outperforms the baseline model by 21.27 points. Distilled from the InfoXLM Large (Chi et al., 2021a) model will further improve the performance by an absolute 2.26-point.

4.5 Ablation Studies

We conduct ablation studies on the test dataset of MASSIVE under the two settings. Table 4 shows the results under the full-dataset setting. Ablating subword sampling will degrade the performance by 1.36 points on the exact match accuracy, where the performance drop comes mainly from the slot filling task, indicating the subword sampling augmentation mainly works on slot filling. Ablating consistency regularization will degrade the performance by 2.19 points on the exact match accuracy. The performances on both intent detection and slot filling tasks are decreased.

The zero-shot setting results are presented in Table 5. It can be observed that when machine translation augmentation is removed, the exact match accuracy drops by 16.68 points, while the performance on intent detection and slot filling are also significantly worse. We also removed the subword sampling augmentation, and the performance is found to have the same trend as in the full-dataset setting. An absolute 1.24-point drop on the exact match accuracy and an absolute 1.75-point drop on slot micro F1 demonstrate that subword sampling is more beneficial for the slot filling task. By removing the consistency regularization, the performance of exact match accuracy will degrade by 2.23 points. The performance shows a significant performance drop on both intent detection and slot filling tasks.

5 Conclusion

We propose to use consistency regularization based on a hybrid data augmentation strategy to improve the performance of multilingual SLU. The proposed method is flexible and can be easily plugged into the fine-tuning process of both the encoder-only model and the Text-to-Text model. The experimental results demonstrate the importance of consistency regularization and the hybrid data augmentation strategy, respectively.

Acknowledgments

This work was supported by the National Key RD Program of China via grant 2020AAA0106501 and the National Natural Science Foundation of China (NSFC) via grant 62236004 and 61976072.

References

Zewen Chi, Li Dong, Furu Wei, Nan Yang, Saksham Singhal, Wenhui Wang, Xia Song, Xian-Ling Mao, Heyan Huang, and Ming Zhou. 2021a. InfoXLM: An information-theoretic framework for cross-lingual language model pre-training. In *Proceedings of the 2021 Conference of the North American Chapter of the Association for Computational Linguistics: Human Language Technologies*, pages 3576–3588, Online. Association for Computational Linguistics.

Zewen Chi, Li Dong, Bo Zheng, Shaohan Huang, Xian-Ling Mao, Heyan Huang, and Furu Wei. 2021b. Improving pretrained cross-lingual language models via self-labeled word alignment. In *Proceedings of the 59th Annual Meeting of the Association for Computational Linguistics and the 11th International Joint Conference on Natural Language Processing (Volume 1: Long Papers)*, pages 3418–3430, Online. Association for Computational Linguistics.

Zewen Chi, Shaohan Huang, Li Dong, Shuming Ma, Bo Zheng, Saksham Singhal, Payal Bajaj, Xia Song, Xian-Ling Mao, Heyan Huang, and Furu Wei. 2022. XLM-E: cross-lingual language model pre-training via ELECTRA. In *Proceedings of the 60th Annual Meeting of the Association for Computational Linguistics (Volume 1: Long Papers), ACL 2022, Dublin, Ireland, May 22-27, 2022*, pages 6170–6182. Association for Computational Linguistics.

Alexis Conneau, Kartikay Khandelwal, Naman Goyal, Vishrav Chaudhary, Guillaume Wenzek, Francisco Guzmán, Edouard Grave, Myle Ott, Luke Zettlemoyer, and Veselin Stoyanov. 2020. Unsupervised cross-lingual representation learning at scale. In *Proceedings of the 58th Annual Meeting of the Association for Computational Linguistics*, pages 8440–8451, Online. Association for Computational Linguistics.

Alexis Conneau and Guillaume Lample. 2019. Cross-lingual language model pretraining. In *Advances in Neural Information Processing Systems*, pages 7057–7067. Curran Associates, Inc.

Jack FitzGerald, Christopher Hench, Charith Peris, Scott Mackie, Kay Rottmann, Ana Sanchez, Aaron Nash, Liam Urbach, Vishesh Kakarala, Richa Singh, Swetha Ranganath, Laurie Crist, Misha Britan, Wouter Leeuwis, Gokhan Tur, and Prem Natarajan. 2022. Massive: A 1m-example multilingual natural language understanding dataset with 51 typologically-diverse languages.

Haoming Jiang, Pengcheng He, Weizhu Chen, Xiaodong Liu, Jianfeng Gao, and Tuo Zhao. 2020. SMART: Robust and efficient fine-tuning for pre-trained natural language models through principled regularized optimization. In *Proceedings of the 58th Annual Meeting of the Association for Computational Linguistics, ACL 2020, Online, July 5-10, 2020*, pages 2177–2190. Association for Computational Linguistics.

Taku Kudo. 2018. Subword regularization: Improving neural network translation models with multiple subword candidates. In *Proceedings of the 56th Annual Meeting of the Association for Computational Linguistics, ACL 2018, Melbourne, Australia, July 15-20, 2018, Volume 1: Long Papers*, pages 66–75. Association for Computational Linguistics.

Taku Kudo and John Richardson. 2018. Sentencepiece: A simple and language independent subword tokenizer and detokenizer for neural text processing. In *Proceedings of the 2018 Conference on Empirical Methods in Natural Language Processing, EMNLP 2018: System Demonstrations, Brussels, Belgium, October 31 - November 4, 2018*, pages 66–71. Association for Computational Linguistics.

Xiaodong Liu, Hao Cheng, Pengcheng He, Weizhu Chen, Yu Wang, Hoifung Poon, and Jianfeng Gao. 2020. Adversarial training for large neural language models. *CoRR*, abs/2004.08994.

Libo Qin, Qiguang Chen, Tianbao Xie, Qixin Li, Jian-Guang Lou, Wanxiang Che, and Min-Yen Kan. 2022. GL-CLeF: A global–local contrastive learning framework for cross-lingual spoken language understanding. In *Proceedings of the 60th Annual Meeting of the Association for Computational Linguistics (Volume 1: Long Papers)*, pages 2677–2686, Dublin, Ireland. Association for Computational Linguistics.

Libo Qin, Minheng Ni, Yue Zhang, and Wanxiang Che. 2020. CoSDA-ML: Multi-lingual code-switching data augmentation for zero-shot cross-lingual NLP. In *Proceedings of the Twenty-Ninth International Joint Conference on Artificial Intelligence, IJCAI 2020*, pages 3853–3860. ijcai.org.

Jasdeep Singh, Bryan McCann, Nitish Shirish Keskar, Caiming Xiong, and Richard Socher. 2019. XLDA: cross-lingual data augmentation for natural language inference and question answering. *CoRR*, abs/1905.11471.

Ashish Vaswani, Noam Shazeer, Niki Parmar, Jakob Uszkoreit, Llion Jones, Aidan N Gomez, Łukasz Kaiser, and Illia Polosukhin. 2017. Attention is all you need. In *Advances in Neural Information Processing Systems*, pages 5998–6008. Curran Associates, Inc.

Linting Xue, Noah Constant, Adam Roberts, Mihir Kale, Rami Al-Rfou, Aditya Siddhant, Aditya Barua, and Colin Raffel. 2021. mt5: A massively multilingual pre-trained text-to-text transformer. In *Proceedings of the 2021 Conference of the North American Chapter of the Association for Computational Linguistics: Human Language Technologies, NAACL-HLT 2021, Online, June 6-11, 2021*, pages 483–498. Association for Computational Linguistics.

Bo Zheng, Li Dong, Shaohan Huang, Saksham Singhal, Wanxiang Che, Ting Liu, Xia Song, and Furu Wei. 2021a. Allocating large vocabulary capacity for cross-lingual language model pre-training. In *Proceedings of the 2021 Conference on Empirical Methods in Natural Language Processing, EMNLP 2021, Virtual Event / Punta Cana, Dominican Republic, 7-11 November, 2021*, pages 3203–3215. Association for Computational Linguistics.

Bo Zheng, Li Dong, Shaohan Huang, Wenhui Wang, Zewen Chi, Saksham Singhal, Wanxiang Che, Ting Liu, Xia Song, and Furu Wei. 2021b. Consistency regularization for cross-lingual fine-tuning. In *Proceedings of the 59th Annual Meeting of the Association for Computational Linguistics and the 11th International Joint Conference on Natural Language Processing (Volume 1: Long Papers)*, pages 3403–3417, Online. Association for Computational Linguistics.

Play música alegre: A Large-Scale Empirical Analysis of Cross-Lingual Phenomena in Voice Assistant Interactions

Donato Crisostomi
Alexa AI, Amazon
Sapienza, University of Rome
doncris@amazon.com

Alessandro Manzotti
Alexa AI, Amazon
manzotti@amazon.com

Enrico Palumbo
Alexa AI, Amazon
palumboe@amazon.com

Davide Bernardi
Alexa AI, Amazon
dvdbe@amazon.com

Sarah Campbell
Alexa AI, Amazon
srh@amazon.com

Shubham Garg
Alexa AI, Amazon
gargshu@amazon.com

Abstract

Cross-lingual phenomena are quite common in informal contexts like social media, where users are likely to mix their native language with English or other languages. However, few studies have focused so far on analyzing cross-lingual interactions in voice-assistant data, which present peculiar features in terms of sentence length, named entities, and use of spoken language. Also, little attention has been posed to European countries, where English is frequently used as a second language. In this paper, we present a large-scale empirical analysis of cross-lingual phenomena (code-mixing, linguistic borrowing, foreign named entities) in the interactions with Alexa in European countries. To do this, we first introduce a general, highly-scalable technique to generate synthetic mixed training data annotated with token-level language labels and we train two neural network models to predict them. We evaluate the models both on the synthetic dataset and on a real dataset of code-switched utterances, showing that the best performance is obtained by a character convolution based model. The results of the analysis highlight different behaviors between countries, having Italy with the highest ratio of cross-lingual utterances and Spain with a marked preference in keeping Spanish words. Our research, paired to the increase of the cross-lingual phenomena in time, motivates further research in developing multilingual Natural Language Understanding (NLU) models, which can naturally deal with cross-lingual interactions.

1 Introduction

The interaction of different languages produces a variety of linguistic phenomena, the most prominent examples being code-switching and lexical borrowing. Code-swiching (CS), or code-mixing[1], refers to the alternation of languages within an utterance or a conversation (Poplack, 2004), while linguistic borrowing occurs when a word is adopted from a language and integrated into another without translation. Examples of these are: (i) "Play música alegre" (ii) "Bravo, that was a great performance", with the former being a case of code-switching and the latter exhibiting lexical borrowing. These phenomena are particularly frequent in bilingual countries, where the local language, called frame-language, is influenced by a second language, which is instead called the mixing-language. This phenomenon is abstracted by the *Matrix Language Frame* model (Poulisse, 1998) in code-switching literature. Common pairs of frame-mixing languages are for example *Spanglish* (Spanish-English) and *Hinglish* (Hindi-English).

Countries for which these phenomena happen usually undergo a broader influence which also permeates their culture, as it happens for example with American artistic production of cinema and music. As a side effect, utterances originated in the frame language are rich in foreign named entities, which contribute to their linguistic heterogeneity. Voice assistants operating in these locales have to face a significant amount of foreign words while being in most cases trained on monolingual corpora, hence posing a severe threat to their performance.

Indeed, the growing interest in multi-lingual models (Devlin et al., 2019; Alexis and Lample, 2019; Conneau et al., 2020) and datasets (FitzGerald et al., 2022; Xu et al., 2020) may help mitigate the problem. We will use in the rest of the paper

[1]We will use the terms code-switching and code-mixing interchangeably, despite they are sometimes used in linguistic literature to denote different phenomena.

Proceedings of the Massively Multilingual Natural Language Understanding Workshop (MMNLU-22), pages 42 - 52
December 7, 2022

the term *cross-lingual* to denote utterances which contain one or more words from a mixing language while belonging to a frame language. These may be caused by any of the mentioned phenomena, *i.e.* code-switching, lexical borrowing and foreign named entities.

A major challenge, both in improving the performances of multilingual models on cross-lingual data and in their overall evaluation, is the scarcity of cross-lingual datasets. Nevertheless, while human annotation is already costly and time-consuming in general, annotating cross-lingual data is made harder by the fact that bilingual annotators are needed for each pair of languages of interest; these may be especially hard to find for less common languages. In particular, while there has been some interest for different kinds of data (*e.g.* social media), voice assistant data, which is the focus of this paper, has been mostly ignored. Although such datasets may be obtained by crowdsourcing, the process would be expensive and time-consuming. This reason leads to the necessity of a procedure to generate synthetic data over several language pairs while providing large-scale datasets. These can be used to train a learning model to infer cross-lingual utterances. The trained model can finally be employed on voice assistants data to detect real cross-lingual utterances.

Our contribution is three-fold: (i) We propose in section 3 a scalable synthetic data generation technique to obtain challenging benchmarks which exhibit a significant ratio of cross-lingual influences. The method is language agnostic and here we employ it on four common European languages (German, French, Italian, Spanish) with English as mixing language. (ii) We compare the performance of different baselines in detecting cross-lingual utterances by solving the more fine-grained task of word level language identification. To validate the generation procedure, we test the models trained on the synthetic distribution over a benchmark dataset obtained through an extremely precise heuristic. (iii) Finally, we analyze in section 6 the phenomenon of cross-lingual influence in a large set of cross-lingual utterances detected using our method on Alexa user queries.

2 Related work

Code switching has received significant interest both in the linguistic literature (Poplack, 2004, 1980; Lipski, 2005; Bhatt and Bolonyai, 2011) and in Natural Language Processing (NLP); (Sitaram et al., 2019) provide a survey of code-switching in NLP. From a linguistic point of view, the two phenomena differ in the fact that the latter occurs in the lexicon, while code-switching mostly regards the utterance-construction level (Muysken, 1995). Despite the apparently different definitions, the two are not always clearly distinct from one another, and may be thought of as lying on a continuum (Sitaram et al., 2019; Bali et al., 2014).

	de	fr	it	es
code switched	31359	5391	6139	4256
non code switched	63944	18491	20100	23744

Table 1: Size of the four benchmark datasets.

Various efforts have been made to collect code-switched annotated data over which to perform core NLP tasks, such as NER (Aguilar et al., 2018; Singh et al., 2018), POS (Vyas et al., 2014; Barman et al., 2016) and ASR (Lyu et al., 2015; Deuchar et al., 2014). Nevertheless, most of the available resources have been gathered from Twitter, and therefore do not resemble the distribution of data encountered by a voice assistant. Few works exist on generating synthetic CS data: in (Pratapa et al., 2018), a synthetic dataset is obtained by applying linguistic theory-based rules, while in (Gupta et al., 2020) an encoder-decoder architecture is used for the generation. These approaches, however, focus on strict code-switching, while we aim to also encompass lexical borrowing and foreign named entities.

The de-facto standard way to infer code-switched utterances is to train models on the task of word-level language identification. Again, existing datasets of code-switched text annotated with word-level language labels have been collected from Twitter (Patro et al., 2017; Maharjan et al., 2015) or Facebook (Barman et al., 2014), leaving conversational data out of the scope. Provided a word-level annotated dataset, any sequence-labeling algorithm can be employed to solve the task. Approaches include conditional random fields (Sikdar and Gambäck, 2016; Shrestha, 2016), recurrent neural networks (Chang and Lin, 2014; Samih et al., 2016) and transfer learning (Aguilar and Solorio, 2020).

3 Data

3.1 Synthetic data generation

As anticipated in section 1, the cost in time and resources of annotating large-scale datasets by crowdsourcing makes synthetic generation the only viable alternative. However, these phenomena show a significant degree of mutability both in time and space (Sitaram et al., 2019), making them elusive to be addressed in a unified manner which is theoretically sound. While some have tried to generate linguistically-correct code-switched data (Pratapa et al., 2018), we trade off a rigorous formulation with a simpler one to deal with all the considered phenomena in a unified manner. Requiring no real cross-lingual (CL) samples, our scalable approach generalizes among any pair of languages. We show that this relaxation does not undermine the effectiveness of the approach by benchmarking a model trained on such generated data over a high-precision CL dataset ("benchmark dataset"). Indeed, our objective is to generate a dataset rich of cross-linguality which can be used to train a model able to detect any CL utterance (code-switching, language borrowing etc.).

The generation follows (Gella et al., 2014), where each utterance can have at most two languages and at most one switching point. While these may not be true in general, they mostly hold in voice assistant data, where utterances are usually short.

Slot switching Our procedure leverages *slot resolution artifacts* which are typically available to conversational agents[2]: these, in fact, need to map entities to actionable items, *e.g.* both '*chapter*', '*section*' and '*paragraph*' are mapped to a coarser entity type which denotes more generally a part of a book. Slot resolution artifacts are usually implemented as human-authored many-to-one maps, where the fine-grained entities are language-specific and the coarser entity type is language-agnostic. The latter can be used as a syntactically safe switching point to obtain cross-lingual utterances. A cross-lingual dataset can be obtained from a chosen monolingual dataset in the frame language by matching instantiations of entity types in the frame-locale utterances and replacing them with random instantiations of the same entity type in the mixing locale. Then, to obtain the token-level language annotations, it is sufficient to assign each switched token to the mixing language. For example, for

(1) "A_{IT} che_{IT} $\text{capitolo}_{\text{IT}}$ sono_{IT} $\text{arrivato}_{\text{IT}}$"

we use the map {capitolo, sezione, paragrafo $\rightarrow$ BOOKSECTION} to obtain the language-agnostic entity 'BOOKSECTION' which contains a set of its instantiations in English (or any other language) {chapter, section, paragraph $\rightarrow$ BOOKSECTION}, allowing us to pick one to produce

(2) "A_{IT} che_{IT} $\text{chapter}_{\text{EN}}$ sono_{IT} $\text{arrivato}_{\text{IT}}$".

We empirically set the mixing probability to 70% after inspecting a subset of utterances. As the mapping from the language-agnostic entities to their instantiations in a chosen language is not univocal, we choose one of the latter at random.

Named-entities switching Nevertheless, slot resolution artifacts only cover specific slots. Another common phenomenon is the use of English words in named entities, such as song names, video names or app names. To obtain a reliable language annotation for named entities, we use a high-precision and low-recall heuristic that checks that each token of the named entity is part of only a specific language dictionary. For instance, when using IT as a frame language and EN as a mixing language, given a song such as "*nel blu dipinto di blu*" we check if '*nel*', '*blu*', '*dipinto*', '*di*', '*blu*' are all part of the IT dictionary and none of them is part of the EN dictionary. Only in that case they are placed in the IT catalog; if the converse happens, they are placed in the EN catalog. Entities for which none of these events happens are not switched. We populate the language-specific catalogs from the data and replace the named entities sampling from either the frame or mixing catalogs of the same entity type (e.g. "Song" $\rightarrow$ sample using the song names catalogs) with a probability proportional to the catalog size. This method creates fairly representative utterances in the context of personal assistants, since we mainly have short sentences with cross-linguality concentrated on named entities and loanwords. While the framework is general and can be used for any pair of languages, we used English as mixing language for the four considered

[2]As an alternative, publicly available resources may also be used: a slot can replaced with a word in the same WordNet synset (Fellbaum, 1998). WordNet has been translated and adapted to many languages, like German, French, Italian, and Spanish (Hamp and Feldweg, 1997; Sagot and Fišer, 2008; Toral et al., 2010; Gonzalez-Agirre et al., 2012).

European languages to mimic the real linguistic phenomenon. We applied our method to manually annotated, de-identified and anonymized Alexa utterances. These span more than two years of data for all the languages considered. Starting from these data we create our cross-lingual data set. We generated four datasets of $\approx 100k$ utterances for the four corresponding locales, each split in training, validation and test with a 80-10-10 ratio. This size was chosen to keep a fairly high variance of the English words present in the utterances.

It is worth to note that we do not require, and hence do not expect, the generated utterances to faithfully resemble the cross-lingual phenomena that we aim to capture. In fact, adhering to the definition of cross-linguality that we outlined in section 1, we more simply aim to generate utterances in a frame language containing one or more words from a mixing language, possibly preserving the original syntax and semantics. If we now consider the set of natural cross-lingual utterances to be a subset of all the possible cross-lingual utterances, we have that a model capable of detecting samples from the former should also be able to detect those from the latter. Given that the set of natural cross-lingual utterances is constrained by the linguistic patterns of the considered phenomena, the subset assumption makes intuitive sense but is not assumed to hold for all distributions. We show, however, that this assumption is valid enough to capture most of the cross-linguality in conversational data, assessing the effectiveness of models trained on the synthetic distribution on a benchmark of real cross-lingual utterances.

3.2 Benchmark dataset

To validate our data generation technique, we need a ground-truth dataset over which to evaluate the proposed models after they have been trained on the generated distribution. Provided that no such dataset exists for conversational data, we take inspiration from (Mendels et al., 2018) to obtain a high-precision set of utterances from de-identified and anonymized live traffic. The approach leverages the idea of anchor words, *i.e.* words belonging specifically to one language among a large pool of languages. Provided anchor words for both the frame and mixing languages, an utterance is code-switched if it contains both an anchor from the frame and one from the mixing language. Analogously to (Mendels et al., 2018), we relax the definition of anchor word by restricting the pool of languages to contain only the mixing language, yielding what are called *weak* anchor words. This is motivated by the fact that most foreign words in the considered frame languages are English, so this relaxation significantly improves the recall while keeping its false positive rate minimum. The set of weak anchor words for the frame language L can be computed as the set difference between its word lexicon V_L and the lexicon of the mixing language $V_{L'}$

$$\text{AnchorSet}(L) = V_L \setminus V_{L'}. \tag{1}$$

The set of weak anchor words for the mixing language can be computed in the symmetric way.

While this procedure has limitations in terms of recall, the obtained set of utterances exhibits almost no false positives. Nevertheless, to obtain a benchmark dataset over which to evaluate both False Positive Rate (FPR) and recall of the trained models, negative samples are also needed. For this we use the set of utterances for which all the words are anchor words of the frame language. As before, although many not code-switched utterances will be this way ignored, the resulting ones will be negative samples with extremely high confidence.

To avoid making assumptions on the ratio of code-switched utterances, the two datasets are kept separated. The one consisting of only code-switched utterances is used to compute the recall, while the one containing only non-code-switched utterances is used to compute the FPR. Table 1 shows the dimensions of the four datasets.

4 Models

We describe in this section the proposed baselines, namely an ad hoc deterministic heuristic and two neural models. These will be trained over the synthetic datasets generated according to section 3 and used to infer real code-switched Alexa utterances.

We consider as baseline a dictionary-based heuristic parameterized by two thresholds t_1 and t_2. The latter deterministically classifies an utterance as code mixed if at least $t_1\%$ of the lemmatized words do not appear in the frame language vocabulary while appearing in the mixing language vocabulary and no more than $t_2\%$ appear in the mixing vocabulary while not belonging to the frame vocabulary. Despite its simplicity, the heuristic allows to arbitrarily trade-off recall and precision by manually tweaking the two parameters.

We then propose two neural models, one character based and the other transformer based. The intuition behind the former is that character-level convolutions (Sitaram et al., 2019) should be able to capture the distinguishing morphological features of the considered languages which are key to the task. In particular, given an input utterance, each word is split in characters and embedded via a trainable embedding layer to obtain $\mathbf{w} \in \mathbb{R}^{l \times d}$, where l is the maximum word length encountered in the data and $d = 50$ is the chosen embedding dimension. The embedded word is then passed through a set of $m = 256$ 1-D convolutional filters with kernel size $k = 3$, yielding a tensor $\in \mathbb{R}^{m \times o}$, where o is given by $(l - k + 1)$. At this point, the maximum is taken along the axis on which the resulting feature maps are stacked, so to have a new word embedding tensor $\mathbf{e} \in \mathbb{R}^{o}$. Three different sets of filters of different kernel sizes are then passed over $\mathbf{e}$, having sizes $3, 4$ and 5 in our implementation. Max pooling over time allows to obtain a fixed-dimension digest for each of the resulting maps, which can be concatenated to form a single tensor to be fed to a bidirectional LSTM along with the rest of the utterance. The latter returns a dynamic representation of the word and its context, which is then mapped to the label space by a standard fully-connected layer. A visual overview of the architecture is given in fig. 1. We will refer to this model as '`CharBased`'. The second proposed neural model leverages multilingual BERT (Devlin et al., 2019) to obtain contextualized embeddings which are then fed to a standard sequence classification pipeline, as can be seen in fig. 2. In details, each word is first tokenized and encoded by the mBERT tokenizer and fed to a pretrained mBERT model along with the whole utterance. The embedding is then provided by the last hidden state of the pretrained model. Since the tokenizer is based on the Wordpiece model (Schuster and Nakajima, 2012), words are often split in subwords: the word '*microfono*' for example would be split in '*micro*' and '*##fono*'. To still obtain word-level predictions, the resulting embeddings are averaged. Utterances are finally fed to a bidirectional LSTM whose output is mapped to the label space again by a fully-connected layer. We will refer to this model as '`BertBased`' in the rest of the paper.

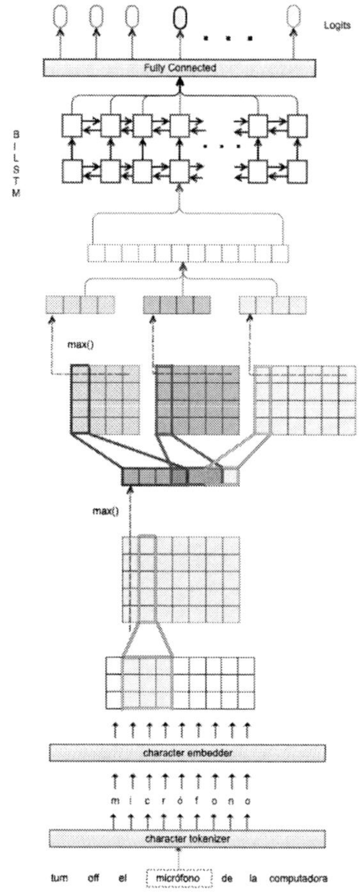

Figure 1: Diagram of the character-convolution-based model.

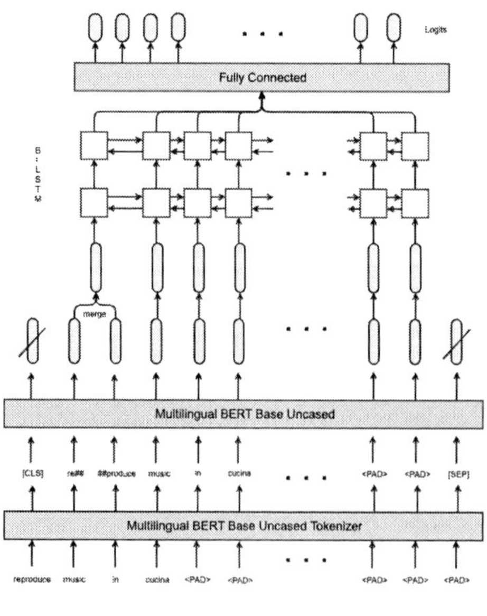

Figure 2: Diagram of the contextual model.

Code-Switching Detection on synthetic data												
	DE			FR			IT			ES		
	F1	prec	recall	F1	prec	recall	F1	prec	recall	F1	prec	recall
`Baseline`	+0%	+0%	+0%	+0%	+0%	+0%	+0%	+0%	+0%	+0%	+0%	+0%
`CharBased`	+25.1%	+35.9%	+9%	+23.9%	+35.7%	+8%	+29.4%	+40.7%	+12.7%	+29.7%	+40.0%	+13.3%
`BertBased`	+26.4%	+37.7%	+10.5%	+25.8%	+36.2%	+11.2%	+30.6%	+40.9%	+15.1%	+31.2%	+42.2%	+14.1%

Table 2: Evaluation results for the task of code-switched utterance detection of the two neural models expressed as relative improvement over the threshold based `Baseline` presented in section 4, performed over a held-out artificially generated test set.

Code-Switching Detection on benchmark data								
	DE		FR		IT		ES	
	recall	FPR	recall	FPR	recall	FPR	recall	FPR
`Baseline`	+0%	+0%	+0%	+0%	+0%	+0%	+0%	+0%
`BertBased`	+7.7%	−22.4%	−1.9%	−48.4%	+2.5%	−46.6%	−0.2%	−35.5%
`CharBased`	+18.4%	−21.3%	+2.9%	−48.7%	+7.9%	−46.2%	+3.4%	−35.4%

Table 3: Evaluation results for the task of code-switched utterance detection of the two neural models expressed as relative improvement over the threshold based `Baseline` presented in section 4, performed over the benchmark dataset obtained as in section 3.

	DE	FR	IT	ES
DE		+16%	−1%	+53%
FR			−15%	+31%
IT				+54%

Table 4: Relative difference in % of utterances containing cross-lingual phenomena by country. Cell ij contains the difference in the ratio of cross-lingual utterances between language i and language j.

5 Evaluation

As can be seen in table 2, the two neural models obtain similar results on a held-out test set generated according to the same procedure presented in section 3, with `BertBased` slightly outperforming the character based model. On the other hand, table 3 shows that the latter obtains the best results on the benchmark dataset, yielding much higher recall while maintaining a low False Positive Rate (FPR). The results are expressed as relative improvements of the two models over the deterministic heuristic introduced in section 4. Precision and recall are given in table 3 because they are computed on two separate datasets to avoid having to pick an arbitrary ratio between code-switched and non-code-switched utterances.

6 Results

Object of this analysis are code-switched utterances detected from real Alexa queries by a model trained on an artificial dataset generated according to section 3. A separate model was trained for each locale versus English, and the inference was made on real data coming from the corresponding locale. As can be seen in table 4, German, French and Italian exhibit similar ratios of cross-lingual utterances, with Italy being the country where they are most common. On the other hand, Spanish shows a remarkably different situation. As shown in fig. 4, this difference is mostly attributable to English words which do not represent named entities: in Spain, people for example do not use '*timer*', '*computer*' or '*film*', as they prefer their Spanish correspondants '*temporizadora*', '*computatora*' and '*pelicula*'. This phenomenon is confirmed in fig. 3, where we see the most common words causing cross-linguality. Figure 3 also shows that the distribution is extremely skewed: for instance, '*timer*' in Italian causes almost the 10% of all the cross-lingual utterances. This phenomenon reflects the underlying distribution of voice assistant utterances, where a set of frequent queries make up for a large part of all of utterances. Finally, we can see in fig. 5 the way cross-lingual utterances are distributed in different domains is common to the different locales. Coherently with the large amount of foreign named entities causing cross-linguality, we can see that most utterances belong to 'Media & Entertainment', which is expected to contain many international artists and song names. '*DeviceControl*' also accounts for a significant part of the utterances; these usually contain commands, like for example '*play*', '*next*', '*stop*' etc., which are traditionally expressed in English even in non-English speaking countries.

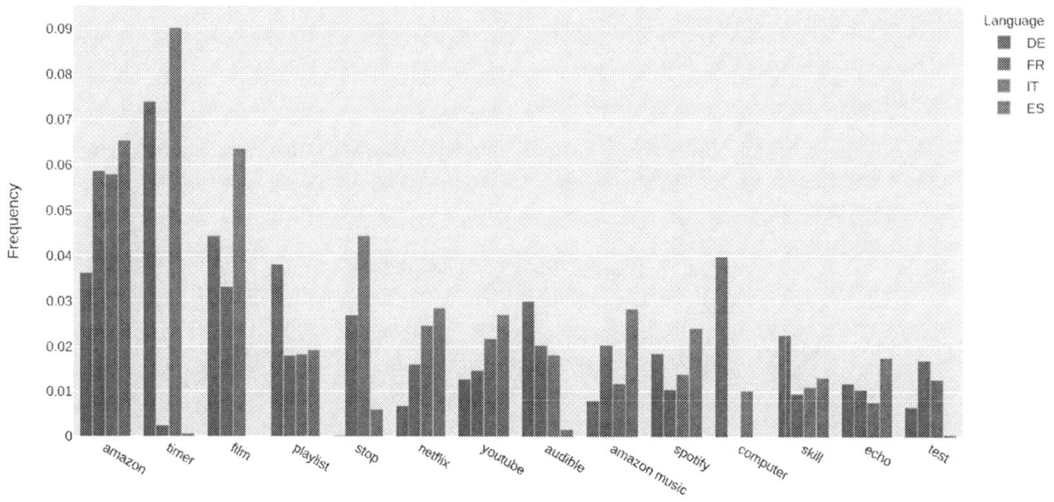

Figure 3: Most common English words used when interacting with Alexa in the four considered locales.

7 Conclusions

In this paper, we have presented a large-scale analysis of the cross-lingual phenomena encountered by voice assistants. We first have proposed an artificial data generation technique, then we have presented two neural models that can be trained on the synthetic data to infer real cross-lingual utterances. Finally, we have employed the top-performing model to infer such utterances from real data. The fact that loanwords and foreign named entities cover most of the found cross-lingual utterances may indicate that code-switching is rare in voice assistants in the considered locales. This may be explained by the fact that users code-switch the most in colloquial situations, while their way of speaking when querying a voice assistant is constrained by its understanding capacity. Nonetheless, multilingual models still have a great opportunity of transfer learning on the large amount of foreign named entities and loanwords that are present in the data. The results show that the use of English words in DE, FR, IT, ES is strongly skewed on popular entities such as '*Amazon*', '*Netflix*', and '*YouTube*', and on specific loanwords such as '*timer*', '*computer*' and '*stop*'. The use of these popular named entities is consistent across locales and the ratio of cross-lingual interactions is similar, except for ES, where users tend to prefer Spanish words to English loanwords. The analysis also shows that most of the mixing words are contained in the 'Media & Entertainment' domains and on named entities such as Service Names, Media names, Item names and Dish names.

As we have explained in section 3, the current generation technique does not aim to model the complex phenomenon of code-switching in a theoretically correct manner. The simplicity of the procedure nevertheless allows it to be repurposed to focus on the latter. An interesting future direction could be to limit the attention to code-switching in the data generation, so that a model trained on that data could be used to collect a code-switched dataset of voice assistant queries. Given the low FPR exhibited by the model, the collected utterances represent an high-quality resource which could in future be used to train generative models to produce better synthetic data, which in turn can be used to train detectors in an iterative manner.

From an architectural prospective, models tackling word-level language identification expressly designed to solve the task of cross-lingual or code-switched detection could benefit from the utterance-level information about their distribution in the dataset. This could encourage the design of a multi-headed model tackling both tasks in an end-to-end approach.

Finally, we aim to expand the set of considered languages to encompass other frame and mixing languages, for example considering Hinglish in India. It might be particularly interesting to compare the obtained results for Spanish with ones obtained over Spanish spoken in the United States and in Mexico, as they may involve more code-switching.

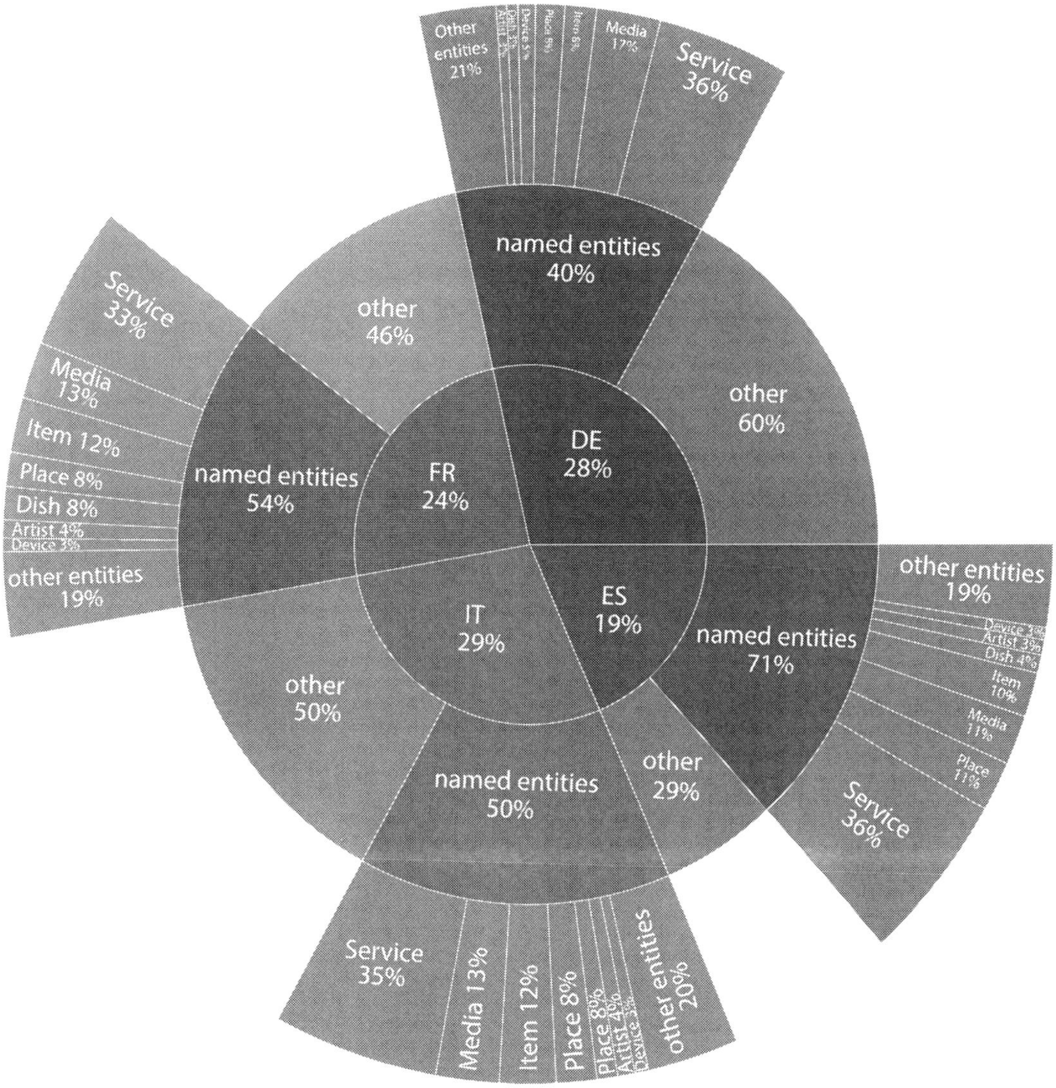

Figure 4: Distribution of a set of $\approx$ 60k cross-lingual utterances.

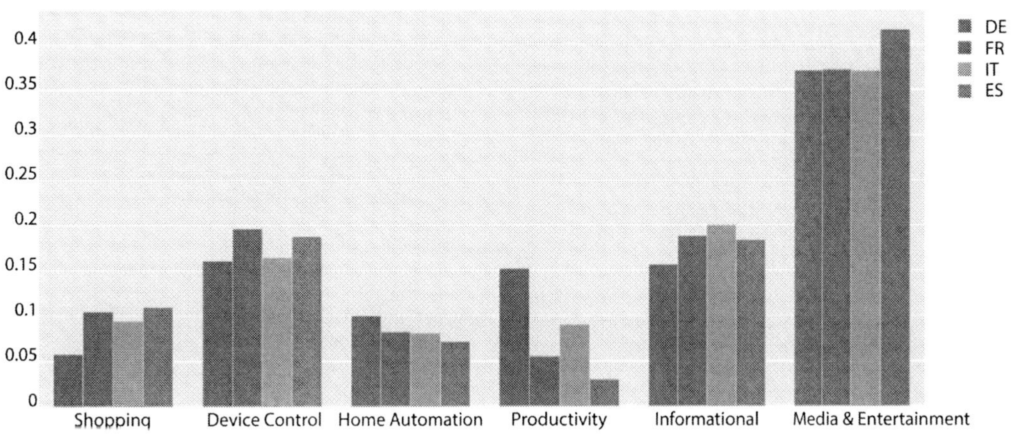

Figure 5: Distribution of domains in cross-linguistic utterances.

8 Limitations

An overall limitation of the work stands from the lack of absolute results, as the latter can only be disclosed as relative improvements over a baseline due to internal policy. As stated in sections 1 and 3, the analysis only regards four European languages (German, French, Italian and Spanish) with English as mixing language. Therefore, while the same approach can be used with different languages, the reported findings only regard the mentioned ones. Moreover, the quality of the generated synthetic data heavily depends on the quality of the slot resolution artifacts presented in section 3. In this work, these artifacts are human-curated according to the highest industry standards, but are subject to IP and hence not publicly accessible. Unfortunately, this also makes the code non disclosable. Finally, as discussed in section 3, the data generation technique may not fully capture the complex linguistic patterns involved in code-switching. We argue that it is however enough to encompass a large quantity of cross-lingual utterances encountered by vocal assistants, and prove it by showing the efficacy of the models trained over synth data in dealing with a high-precision benchmark dataset of real cross-lingual utterances.

References

Gustavo Aguilar, Fahad AlGhamdi, Victor Soto, Mona Diab, Julia Hirschberg, and Thamar Solorio. 2018. Named entity recognition on code-switched data: Overview of the CALCS 2018 shared task. In *Proceedings of the Third Workshop on Computational Approaches to Linguistic Code-Switching*, pages 138–147, Melbourne, Australia. Association for Computational Linguistics.

Gustavo Aguilar and Thamar Solorio. 2020. From English to code-switching: Transfer learning with strong morphological clues. In *Proceedings of the 58th Annual Meeting of the Association for Computational Linguistics*, pages 8033–8044, Online. Association for Computational Linguistics.

Conneau Alexis and Guillaume Lample. 2019. Cross-lingual language model pretraining. In *Advances in Neural Information Processing Systems 32: Annual Conference on Neural Information Processing Systems 2019, NeurIPS 2019, December 8-14, 2019, Vancouver, BC, Canada*, pages 7057–7067.

Kalika Bali, Jatin Sharma, Monojit Choudhury, and Yogarshi Vyas. 2014. "I am borrowing ya mixing ?" an analysis of English-Hindi code mixing in Facebook. In *Proceedings of the First Workshop on Computational Approaches to Code Switching*, pages 116–126, Doha, Qatar. Association for Computational Linguistics.

Utsab Barman, Amitava Das, Joachim Wagner, and Jennifer Foster. 2014. Code mixing: A challenge for language identification in the language of social media. In *Proceedings of the First Workshop on Computational Approaches to Code Switching*, pages 13–23, Doha, Qatar. Association for Computational Linguistics.

Utsab Barman, Joachim Wagner, and Jennifer Foster. 2016. Part-of-speech tagging of code-mixed social media content: Pipeline, stacking and joint modelling. In *Proceedings of the Second Workshop on Computational Approaches to Code Switching*, pages 30–39, Austin, Texas. Association for Computational Linguistics.

Rakesh M. Bhatt and Agnes Bolonyai. 2011. Code-switching and the optimal grammar of bilingual language use. *Bilingualism: Language and Cognition*, 14(4):522–546.

Joseph Chee Chang and Chu-Cheng Lin. 2014. Recurrent-neural-network for language detection on twitter code-switching corpus. *CoRR*, abs/1412.4314.

Alexis Conneau, Kartikay Khandelwal, Naman Goyal, Vishrav Chaudhary, Guillaume Wenzek, Francisco Guzmán, Edouard Grave, Myle Ott, Luke Zettlemoyer, and Veselin Stoyanov. 2020. Unsupervised cross-lingual representation learning at scale. In *Proceedings of the 58th Annual Meeting of the Association for Computational Linguistics*, pages 8440–8451, Online. Association for Computational Linguistics.

Margaret Deuchar, Peredur Davies, Jon Russell Herring, M. Carmen Parafita Couto, and Diana Carter. 2014. *5. Building Bilingual Corpora*, pages 93–110. Multilingual Matters.

Jacob Devlin, Ming-Wei Chang, Kenton Lee, and Kristina Toutanova. 2019. BERT: Pre-training of deep bidirectional transformers for language understanding. In *Proceedings of the 2019 Conference of the North American Chapter of the Association for Computational Linguistics: Human Language Technologies, Volume 1 (Long and Short Papers)*, pages 4171–4186, Minneapolis, Minnesota. Association for Computational Linguistics.

Christiane Fellbaum. 1998. *WordNet: An Electronic Lexical Database*. Bradford Books.

Jack FitzGerald, Christopher Hench, Charith Peris, Scott Mackie, Kay Rottmann, Ana Sanchez, Aaron Nash, Liam Urbach, Vishesh Kakarala, Richa Singh, Swetha Ranganath, Laurie Crist, Misha Britan, Wouter Leeuwis, Gokhan Tur, and Prem Natarajan. 2022. Massive: A 1m-example multilingual natural language understanding dataset with 51 typologically-diverse languages.

Spandana Gella, Kalika Bali, and Monojit Choudhury. 2014. "ye word kis lang ka hai bhai?" testing the limits of word level language identification. In *Proceedings of the 11th International Conference on Natural Language Processing*, pages 368–377, Goa, India. NLP Association of India.

Aitor Gonzalez-Agirre, Egoitz Laparra, and German Rigau. 2012. Multilingual central repository version 3.0: upgrading a very large lexical knowledge base. In *Proceedings of the 6th Global WordNet Conference (GWC 2012)*, Matsue.

Deepak Gupta, Asif Ekbal, and Pushpak Bhattacharyya. 2020. A semi-supervised approach to generate the code-mixed text using pre-trained encoder and transfer learning. In *Findings of the Association for Computational Linguistics: EMNLP 2020*, pages 2267–2280, Online. Association for Computational Linguistics.

Birgit Hamp and Helmut Feldweg. 1997. GermaNet - a lexical-semantic net for German. In *Automatic Information Extraction and Building of Lexical Semantic Resources for NLP Applications*.

John M. Lipski. 2005. Code-switching or borrowing? no sé so no puedo decir, you know.

Dau-Cheng Lyu, Tien-Ping Tan, Eng-Siong Chng, and Haizhou Li. 2015. Mandarin–english code-switching speech corpus in south-east asia: Seame. *Language Resources and Evaluation*, 49(3):581–600.

Suraj Maharjan, Elizabeth Blair, Steven Bethard, and Thamar Solorio. 2015. Developing language-tagged corpora for code-switching tweets. In *Proceedings of The 9th Linguistic Annotation Workshop*, pages 72–84, Denver, Colorado, USA. Association for Computational Linguistics.

Gideon Mendels, Victor Soto, Aaron Jaech, and Julia Hirschberg. 2018. Collecting code-switched data from social media. In *Proceedings of the Eleventh International Conference on Language Resources and Evaluation (LREC 2018)*, Miyazaki, Japan. European Language Resources Association (ELRA).

Pieter Muysken. 1995. *Code-switching and grammatical theory*, page 177–198. Cambridge University Press.

Jasabanta Patro, Bidisha Samanta, Saurabh Singh, Abhipsa Basu, Prithwish Mukherjee, Monojit Choudhury, and Animesh Mukherjee. 2017. All that is English may be Hindi: Enhancing language identification through automatic ranking of the likeliness of word borrowing in social media. In *Proceedings of the 2017 Conference on Empirical Methods in Natural Language Processing*, pages 2264–2274, Copenhagen, Denmark. Association for Computational Linguistics.

Shana Poplack. 1980. Sometimes i'll start a sentence in spanish y termino en espaÑol: toward a typology of code-switching1. 18(7-8):581–618.

Shana Poplack. 2004. *Code-Switching*, pages 589–596.

Nanda Poulisse. 1998. Duelling languages: Grammatical structure in codeswitching. *International Journal of Bilingualism*, 2(3):377–380.

Adithya Pratapa, Gayatri Bhat, Monojit Choudhury, Sunayana Sitaram, Sandipan Dandapat, and Kalika Bali. 2018. Language modeling for code-mixing: The role of linguistic theory based synthetic data. In *Proceedings of the 56th Annual Meeting of the Association for Computational Linguistics (Volume 1: Long Papers)*, pages 1543–1553, Melbourne, Australia. Association for Computational Linguistics.

Benoît Sagot and Darja Fišer. 2008. Building a free French wordnet from multilingual resources. In *Proceedings of the Sixth International Language Resources and Evaluation (LREC'08)*, Marrakech, Morocco.

Younes Samih, Suraj Maharjan, Mohammed Attia, Laura Kallmeyer, and Thamar Solorio. 2016. Multilingual code-switching identification via LSTM recurrent neural networks. In *Proceedings of the Second Workshop on Computational Approaches to Code Switching*, pages 50–59, Austin, Texas. Association for Computational Linguistics.

Mike Schuster and Kaisuke Nakajima. 2012. Japanese and korean voice search. In *2012 IEEE International Conference on Acoustics, Speech and Signal Processing (ICASSP)*, pages 5149–5152.

Prajwol Shrestha. 2016. Codeswitching detection via lexical features in conditional random fields. In *Proceedings of the Second Workshop on Computational Approaches to Code Switching*, pages 121–126, Austin, Texas. Association for Computational Linguistics.

Utpal Kumar Sikdar and Björn Gambäck. 2016. Language identification in code-switched text using conditional random fields and babelnet. In *Proceedings of the Second Workshop on Computational Approaches to Code Switching*, pages 127–131, Austin, Texas. Association for Computational Linguistics.

Vinay Singh, Deepanshu Vijay, Syed Sarfaraz Akhtar, and Manish Shrivastava. 2018. Named entity recognition for Hindi-English code-mixed social media text. In *Proceedings of the Seventh Named Entities Workshop*, pages 27–35, Melbourne, Australia. Association for Computational Linguistics.

Sunayana Sitaram, Khyathi Raghavi Chandu, Sai Krishna Rallabandi, and Alan W. Black. 2019. A survey of code-switched speech and language processing. *CoRR*, abs/1904.00784.

Antonio Toral, Stefania Bracale, Monica Monachini, and Claudia Soria. 2010. Rejuvenating the italian wordnet: upgrading, standardising, extending. In *Proceedings of the 5th International Conference of the Global WordNet Association (GWC-2010)*, Mumbai.

Yogarshi Vyas, Spandana Gella, Jatin Sharma, Kalika Bali, and Monojit Choudhury. 2014. POS tagging of English-Hindi code-mixed social media content. In *Proceedings of the 2014 Conference on Empirical Methods in Natural Language Processing (EMNLP)*, pages 974–979, Doha, Qatar. Association for Computational Linguistics.

Weijia Xu, Batool Haider, and Saab Mansour. 2020. End-to-end slot alignment and recognition for cross-lingual NLU. In *Proceedings of the 2020 Conference on Empirical Methods in Natural Language Processing (EMNLP)*, pages 5052–5063, Online. Association for Computational Linguistics.

Zero-Shot Cross-Lingual Sequence Tagging as Seq2Seq Generation for Joint Intent Classification and Slot Filling

Fei Wang[*]**, Kuan-Hao Huang**[*]**, Anoop Kumar, Aram Galstyan, Greg Ver Steeg, Kai-Wei Chang**
Amazon Alexa AI

Abstract

The joint intent classification and slot filling task seeks to detect the intent of an utterance and extract its semantic concepts. In the zero-shot cross-lingual setting, a model is trained on a source language and then transferred to other target languages through multi-lingual representations without additional training data. While prior studies show that pre-trained multilingual sequence-to-sequence (Seq2Seq) models can facilitate zero-shot transfer, there is little understanding on how to design the output template for the joint prediction tasks. In this paper, we examine three aspects of the output template – (1) label mapping, (2) task dependency, and (3) word order. Experiments on the MASSIVE dataset consisting of 51 languages show that our output template significantly improves the performance of pre-trained cross-lingual language models.

1 Introduction

The joint intent classification and slot filling task is crucial for goal-oriented dialogue systems, seeking to detect the intent of an utterance and extract semantic concepts. This task has been widely studied in the literature (Hakkani-Tür et al., 2016; Zhang and Wang, 2016; Goo et al., 2018). However, due to the difficulty of collecting and annotating large data sets, most studies focus on only a few high-resource languages (e.g., English). To broaden the language coverage of models, zero-shot cross-lingual transfer technique has been proposed (Xu et al., 2020; Li et al., 2021; FitzGerald et al., 2022). Under the zero-shot cross-lingual setting, models are trained on a source language (e.g., English) with sufficient annotated training data and transfer to other target languages.

In particular, recently, FitzGerald et al. (2022) show that pre-trained generative cross-lingual language models (XLMs) (Liu et al., 2020; Xue et al., 2021) can be applied to the joint intent classification and slot filling task. They formulate the joint task as sequence-to-sequence (Seq2Seq) generation, where the model generates the slot label for each word and the intent label for the utterance in a sequential manner based on an output template. However, the design of the output template is usually ad hoc and there is lack of understanding on how different template designs affect the performance of the zero-shot transfer. For example, in Fig. 1, the intent label "alarm set" can be represented by "set alarm". We found that the change of the surface form of the label significantly affects model's performance.

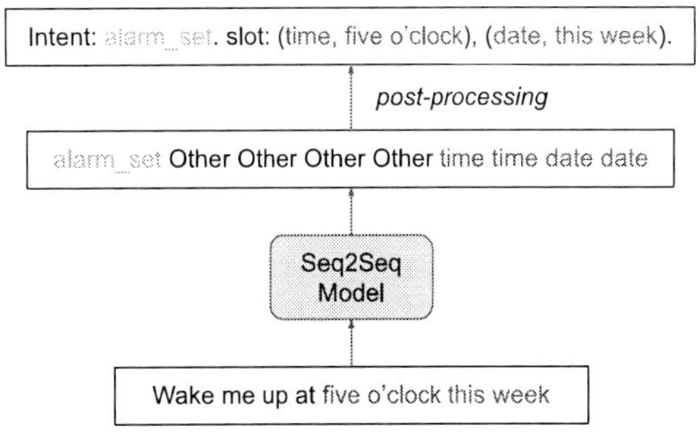

Figure 1: Illustration of Seq2Seq generation for the joint intent classification and slot filling task. Given an input on the bottom, the Seq2Seq model generates the output sequence based on a template – the template forces the model first output the intent label and then slot label of each word in the input sentence. Based on the template, a post-processing step translates the output sequence into structured labels for the task.

In this paper, we examine three aspects in the design of output template, i.e. label mapping, task dependency, and word order. We found that all these aspects have significant influence on model performance. First, based on our observation on label mapping, we propose a concise hierarchical label mapping that leads to a better performance than the default label mapping used in annotation. Second, we observed that generating the intent label before

[*]This work is done during Fei Wang and Kuan-Hao's internship at Amazon.

Proceedings of the Massively Multilingual Natural Language Understanding Workshop (MMNLU-22), pages 53 - 61
December 7, 2022 ©2022 Association for Computational Linguistics

the slots labels leads to better performance. This is because intent classification is a relatively simpler task compared to slot filling and thus the correct task order prevents error propagation. Finally, we found that word shuffle improves the diversity of data and therefore leads to better intent accuracy.

Experiments on the MASSIVE dataset (FitzGerald et al., 2022) consisting of 51 languages demonstrate that our proposed template design can significantly improve the performance of pre-trained generative XLMs on this joint task in the zero-shot transfer setting. We also provide detailed ablation studies and discussion. We intend to release the source code for reproducing our experiments upon paper acceptance.

2 Method

We first provide an overview of the Seq2Seq generation method with pretrained generative XLMs for the joint intent classificaiton and slot filling task. Then, we discuss the design of output template for Seq2Seq generation.

2.1 Seq2Seq for the Joint Prediction Task

Following FitzGerald et al. (2022), we formulate the joint task as Seq2Seq generation, and adopt pretrained XLMs for this task. In this way, we can take advantage of the rich cross-lingual knowledge possessed in the pre-trained XLMs. As shown in Fig. 1, the model generates the intent label for the utterance and the slot label for each word in a sequential manner based on an output template. We insert `Annotate:` at the beginning of the input sequence to indicate the task type. We also insert word separators to indicate the tokens belonging to each word, as we want to generate word-level slot labels. We then use the following objective function to fine-tune the text generation model:

$$\mathcal{L} = -\frac{1}{|\mathbf{y}|}\sum_{k=1}^{|\mathbf{y}|}\log p(y_k|\mathbf{y}_{<k}, \mathbf{x}), \quad (1)$$

where $\mathbf{x}$ is the input utterance sequence and $\mathbf{y}$ is the output label sequence. We provide the ground-truth $\mathbf{y}$ during training. In the following, we explore three aspects of the design of the output template.

2.2 Label Mapping

The first aspect we examine is the label mapping, i.e. the surface form of the labels. When annotating data, the annotators are given a set of output labels. The choice of vocabulary for these labels is often arbitrary as long as the human annotators understand the meaning. However, for a Seq2Seq model, different surface forms of the output labels may lead to different performance even though they are synonyms. For example, the intent label `iot_wemo_on` could be difficult for a fine-tuned XLMs to understand and transfer to other languages. Moreover, labels hold hierarchical relations. For example, some intent labels may belong to the same scenario and some slot labels belong to the same intent. By rephrasing the output labels and leveraging their relations, the model performance can greatly improve.

We propose a concise and hierarchical label mapping based on these observations. For labels belonging to the same scenario or intent, we add the same prefix to them. Some slot labels may belong to multiple scenarios or intents, so we do not add any prefix to them. For example, both `email_folder` and `email_address` belongs to the same scenario so we give them the same prefix `email`, while `time` belongs to multiple scenarios, so we do not give it a prefix. We also remove or replace the redundant and rare words in the labels (e.g. `wemo` in `iot_wemo_on`).

2.3 Task Dependency

The second aspect is the dependency between intent classification and slot filling. In Seq2Seq decoding, the label y_k is conditioned on previously generated tokens $\mathbf{y}_{<k}$. When solving the joint task, the label of one task serves as the condition to generate the label of the other task. Due to this, the later task may benefit from the labels of the former task, but may also suffer from inaccurate predictions of the former task (i.e. error propagation).In particular, we consider two different orders: (1) intent labels before slot labels, and (2) slot labels before intent labels.

2.4 Word Order

Prior works show that reducing word order information in sequence labeling can improve cross-lingual transferability (Ahmad et al., 2019; Liu et al., 2021). This is mainly due to different languages have different word orders (e.g., some languages present adjectives before nouns and some have reverse order), which cause a misalignment in language transfer. In Seq2Seq decoding, changing word orders results in different label order. To make the model more robust on different word orders, we augment the training data by shuffling the utterances and their corresponding labels. However, different from prior works, we shuffle the utterances at the segment level, where words belonging to the same slot is considered as one segment and adjacent words that do not belong to any slot are considered another segment.

3 Experiment

In this section, we evaluate our approach on a massive number of target languages.

3.1 Setup

Dataset. We adopt the MASSIVE (FitzGerald et al., 2022) dataset, which consists of 51 languages, 18 domains, 60 intents, 55 slots and 19,521 utterance per language. We use English data for training and development, data in all the other languages are used for testing.[1] We report the intent accuracy, micro-average slot F1 and exact match accuracy.[2]

Baseline. We compare our method with both classification based and generation based methods. The classification method based on XLM-R (Conneau et al., 2020) formulates the joint task as sequence classification and sequence tagging. Two classification heads are added on top of the pre-trained language model. The generation method based on mT5 (Xue et al., 2021) generates the tag of each word and the intent label in a sequence-to-sequence manner. Following FitzGerald et al. (2022), we use the base version of pre-trained models.

Implementation Details. We evaluate our method based on mT5 with the original model-related hyper-parameters. We follow the hyper-parameters of FitzGerald et al. (2022) for training, except for batch size, learning rate and epochs, which we set to 96, 5e-5 and 200, respectively. We investigate three design choices listed in Sec. 2. We found that better label mapping and task dependency significantly improves the model performance. However, while input shuffle improves intent accuracy, the slot F1 and exact match performance drop. In the following, we will first compare our best model (w/ label mapping and w/ task dependency) with the current state-of-the-art approach, then we will provide detailed ablation study.

[1] This setting is more strict than FitzGerald et al. (2022)'s, where they use data in target languages for development.

[2] Exact match means both the intent and slots are correct.

3.2 Results

Tab. 1 shows the overall model performance on MASSIVE. In comparison with the vanilla mT5, our proposed techniques improve average intent accuracy by 1.5%, average slot F1 by 10.7% and average exact match accuracy by 5.2%. It also changes the highest performing languages in terms of slot F1 and exact match accuracy. These results indicate that task order, label mapping, and other key components in text generation have significant influence on model performance when performing sequence tagging in a sequence-to-sequence generation manner. The key differences between XLM-R based and mT5 based methods are that the latter ones use pre-trained token embeddings as labels and generate each label conditioned on previously generated labels. The vanilla mT5 performs much worse than the prior SOTA method, XLM-R, on all metrics. However, our method based on mT5 achieves better performance than XLM-R in terms of average slot F1 (+5.0%) and exact match accuracy (+2.2%). The failure of vanilla mT5 further shows the importance of well-designed inputs and outputs. The lowest performing language is consistent to be Japanese in all methods. Our method improves slot F1 and exact match accuracy of the lowest performing language.

Performance on Language Characteristics. We further analyze the model performance on different language characteristics. As shown in Fig. 2, our method performs better than vanilla mT5 on 49 out of 50 languages, indicating it can improve the cross-lingual transferability on massive target languages. Norwegian is the only language on which our method performs slightly worse. We provide detailed model performance on 9 language characteristics in Appx. §A, where the languages are split into 3 to 28 groups by each characteristic. Our method improves the performance of all language groups, except for Lolo-Burmese subdivision and Burmese script which contain only Norwegian. We observe that it is difficult to improve model performance on the Japonic and Sino-Tibetan language families when using English as source language. Similarly, a prior work (Malkin et al., 2022) also shows that English may not be an optimal pretraining language in cross-lingual transfer. We leave finding the best general source language for finetuning zero-shot models for future work.

Ablation Study. We also investigate the effective-

Model	Intent Acc (%) High	Intent Acc (%) Low	Intent Acc (%) Avg	Slot F1 (%) High	Slot F1 (%) Low	Slot F1 (%) Avg	Exact Match Acc (%) High	Exact Match Acc (%) Low	Exact Match Acc (%) Avg
mT5	79.9 ± 1.4 nl-NL	25.7 ± 1.6 ja-JP	62.9 ± 0.2	64.3 ± 0.7 de-DE	13.9 ± 0.3 ja-JP	44.8 ± 0.1	53.2 ± 1.8 sv-SE	9.4 ± 1.0 ja-JP	34.7 ± 0.2
XLM-R	85.2 ± 1.3 sv-SE	44.8 ± 1.8 ja-JP	**70.6** ± 0.2	68.4 ± 0.7 sv-SE	15.4 ± 0.3 ja-JP	50.3 ± 0.1	57.9 ± 1.8 sv-SE	9.8 ± 1.1 ja-JP	38.7 ± 0.2
mT5*	80.6 ± 0.7 nl-NL	32.1 ± 0.9 ja-JP	64.8 ± 0.1	63.9 ± 0.3 de-DE	14.7 ± 0.2 ja-JP	44.6 ± 0.1	54.1 ± 0.9 sv-SE	10.1 ± 0.6 ja-JP	35.7 ± 0.1
OURS	80.8 ± 0.7 nl-NL	24.6 ± 0.8 ja-JP	66.3 ± 0.1	71.6 ± 0.5 th-TH	19.6 ± 0.2 ja-JP	**55.3** ± 0.1	57.4 ± 0.9 nl-NL	10.2 ± 0.5 ja-JP	**40.9** ± 0.1

Table 1: Zero-shot cross-lingual results on MASSIVE. We report intent accuracy, micro-averaged slot F1 score, and exact match accuracy of highest language, lowest language and average of all target languages. Best average scores are in bold. Intervals for 95% confidence are given assuming normal distributions. *: results reproduced by us. Other baseline results are copied from the original paper.

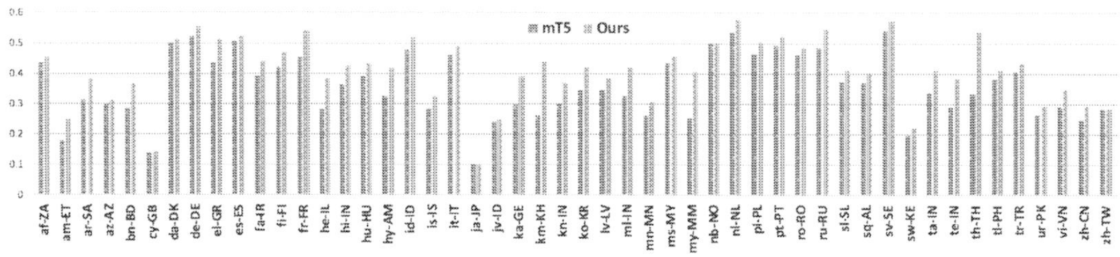

Figure 2: Exact match accuracy of all target languages. Our method performs better than vanilla mT5 on 49 out of 50 languages.

Method	Intent Acc	Slot F1	Exact Match
OURS	66.3	55.3	40.9
OURS w/ default label	67.0	52.5	39.7
OURS w/ slot first	64.6	57.5	37.1
OURS w/ input shuffle	67.5	52.6	40.1

Table 2: Ablation study. All intervals for 95% confidence are within ± 0.1%.

ness of each proposed technique as shown in Tab. 2. Input shuffle increases the diversity of inputs and help the model to avoid overfitting English syntax. Results show that it can improve the intent accuracy (+1.2%); however, it hinders predicting slots accurately. Concise and hierarchical label mapping improves the slot F1 significantly (+2.8%). Task order also plays an important role. Generating slot labels before the intent label for each utterance leads to worse intent accuracy (-1.7%) but better slot F1 (+2.2%). We observe the subtask performance is better when the model generates the subtask labels first.

4 Related Work

Zero-shot cross-lingual joint intent classification and slot filling is crucial for developing goal-oriented dialogue systems for massive languages with less manually annotation (Upadhyay et al., 2018; Li et al., 2021; FitzGerald et al., 2022). Prior works on this joint task can be summarized into two lines. The first line follows a strict zero-shot setting, where only the data in source languages are used for training (Xu et al., 2020; Li et al., 2021; FitzGerald et al., 2022). The second line uses additional data consisting of words or utterances in target languages for training, where the additional data can be annotated data in target languages or synthetic data by code-switching and automatic translation (Upadhyay et al., 2018; Schuster et al., 2019; Krishnan et al., 2021).

Our work follows the strict zero-shot setting. Prior works either formulate the joint task as sequence tagging and applies pretrained cross-lingual encoders (Pires et al., 2019; Conneau et al., 2020) to solve it (Xu et al., 2020; Li et al., 2021; FitzGerald et al., 2022), or formulate it as Seq2Seq generation and applies pretrained generative XLMs to solve it (FitzGerald et al., 2022). Our work analyzes important variables in the output format of the Seq2Seq method.

5 Conclusion

In this paper, we examine three variables of output format in Seq2Seq generation for zero-shot cross-lingual joint intent classification and slot filling. Experiments on the MASSIVE dataset consisting of 51 languages show that all the variables have significant influence on model performance. Specifically, the output format should use a concise and hierarchical label mapping, and consider the label dependency carefully.

Acknowledgement

The authors thank the anonymous reviewers for their valuable comments. They also thank I-Hung Hsu and Muhao Chen for their feedback.

Limitation

In this paper, we analyze three aspects of the design of output format in Seq2Seq generation for zero-shot transfer. There are other factors (e.g., decoding strategy) may also influence the model performance and its transferablity. Besides, this paper focuses on the *output* template of Seq2Seq generation in the cross-lingual transfer setting. We do not consider and compare with other techniques such as data augmentation methods for zero-shot cross-lingual transfer (e.g., code-switching (Qin et al., 2021) and robust training (Huang et al., 2021)).

References

Wasi Ahmad, Zhisong Zhang, Xuezhe Ma, Eduard Hovy, Kai-Wei Chang, and Nanyun Peng. 2019. On difficulties of cross-lingual transfer with order differences: A case study on dependency parsing. In *Proceedings of the 2019 Conference of the North American Chapter of the Association for Computational Linguistics: Human Language Technologies, Volume 1 (Long and Short Papers)*, pages 2440–2452.

Alexis Conneau, Kartikay Khandelwal, Naman Goyal, Vishrav Chaudhary, Guillaume Wenzek, Francisco Guzmán, Édouard Grave, Myle Ott, Luke Zettlemoyer, and Veselin Stoyanov. 2020. Unsupervised cross-lingual representation learning at scale. In *Proceedings of the 58th Annual Meeting of the Association for Computational Linguistics*, pages 8440–8451.

Jack FitzGerald, Christopher Hench, Charith Peris, Scott Mackie, Kay Rottmann, Ana Sanchez, Aaron Nash, Liam Urbach, Vishesh Kakarala, Richa Singh, et al. 2022. Massive: A 1m-example multilingual natural language understanding dataset with 51 typologically-diverse languages. *arXiv preprint arXiv:2204.08582*.

Chih-Wen Goo, Guang Gao, Yun-Kai Hsu, Chih-Li Huo, Tsung-Chieh Chen, Keng-Wei Hsu, and Yun-Nung Chen. 2018. Slot-gated modeling for joint slot filling and intent prediction. In *Proceedings of the 2018 Conference of the North American Chapter of the Association for Computational Linguistics: Human Language Technologies, Volume 2 (Short Papers)*, pages 753–757.

Dilek Hakkani-Tür, Gokhan Tur, Asli Celikyilmaz, Yun-Nung Chen, Jianfeng Gao, Li Deng, and Ye-Yi Wang. 2016. Multi-Domain Joint Semantic Frame Parsing Using Bi-Directional RNN-LSTM. In *Proc. Interspeech 2016*, pages 715–719.

Kuan-Hao Huang, Wasi Ahmad, Nanyun Peng, and Kai-Wei Chang. 2021. Improving zero-shot cross-lingual transfer learning via robust training. In *Proceedings of the 2021 Conference on Empirical Methods in Natural Language Processing*, pages 1684–1697.

Jitin Krishnan, Antonios Anastasopoulos, Hemant Purohit, and Huzefa Rangwala. 2021. Multilingual code-switching for zero-shot cross-lingual intent prediction and slot filling. In *Proceedings of the 1st Workshop on Multilingual Representation Learning*, pages 211–223.

Haoran Li, Abhinav Arora, Shuohui Chen, Anchit Gupta, Sonal Gupta, and Yashar Mehdad. 2021. Mtop: A comprehensive multilingual task-oriented semantic parsing benchmark. In *Proceedings of the 16th Conference of the European Chapter of the Association for Computational Linguistics: Main Volume*, pages 2950–2962.

Yinhan Liu, Jiatao Gu, Naman Goyal, Xian Li, Sergey Edunov, Marjan Ghazvininejad, Mike Lewis, and Luke Zettlemoyer. 2020. Multilingual denoising pre-training for neural machine translation. *Transactions of the Association for Computational Linguistics*, 8:726–742.

Zihan Liu, Genta I Winata, Samuel Cahyawijaya, Andrea Madotto, Zhaojiang Lin, and Pascale Fung. 2021. On the importance of word order information in cross-lingual sequence labeling. In *Proceedings of the AAAI Conference on Artificial Intelligence*, volume 35, pages 13461–13469.

Dan Malkin, Tomasz Limisiewicz, and Gabriel Stanovsky. 2022. A balanced data approach for evaluating cross-lingual transfer: Mapping the linguistic blood bank. In *Proceedings of the 2022 Conference of the North American Chapter of the Association for Computational Linguistics: Human Language Technologies*, pages 4903–4915, Seattle, United States. Association for Computational Linguistics.

Telmo Pires, Eva Schlinger, and Dan Garrette. 2019. How multilingual is multilingual bert? In *Proceedings of the 57th Annual Meeting of the Association for Computational Linguistics*, pages 4996–5001.

Libo Qin, Minheng Ni, Yue Zhang, and Wanxiang Che. 2021. Cosda-ml: multi-lingual code-switching data augmentation for zero-shot cross-lingual nlp. In *Proceedings of the Twenty-Ninth International Conference on International Joint Conferences on Artificial Intelligence*, pages 3853–3860.

Sebastian Schuster, Sonal Gupta, Rushin Shah, and Mike Lewis. 2019. Cross-lingual transfer learning for multilingual task oriented dialog. In *Proceedings of the 2019 Conference of the North American Chapter of the Association for Computational Linguistics: Human Language Technologies, Volume 1 (Long and Short Papers)*, pages 3795–3805.

Shyam Upadhyay, Manaal Faruqui, Gokhan Tür, Hakkani-Tür Dilek, and Larry Heck. 2018. (almost) zero-shot cross-lingual spoken language understanding. In *2018 IEEE International Conference on Acoustics, Speech and Signal Processing (ICASSP)*, pages 6034–6038. IEEE.

Weijia Xu, Batool Haider, and Saab Mansour. 2020. End-to-end slot alignment and recognition for cross-lingual nlu. In *Proceedings of the 2020 Conference on Empirical Methods in Natural Language Processing (EMNLP)*, pages 5052–5063.

Linting Xue, Noah Constant, Adam Roberts, Mihir Kale, Rami Al-Rfou, Aditya Siddhant, Aditya Barua, and Colin Raffel. 2021. mt5: A massively multilingual pre-trained text-to-text transformer. In *Proceedings of the 2021 Conference of the North American Chapter of the Association for Computational Linguistics: Human Language Technologies*, pages 483–498.

Xiaodong Zhang and Houfeng Wang. 2016. A joint model of intent determination and slot filling for spoken language understanding. In *Proceedings of the Twenty-Fifth International Joint Conference on Artificial Intelligence*, pages 2993–2999.

A Model performance on language characteristics

We compare the performance of vanilla mT5 and our method on different language characteristics, including script (Fig. 3), subdivision (Fig. 4), family (Fig. 5), order (Fig. 6), politeness (Fig. 7), imperative morphology (Fig. 8), imperative hortative (Fig. 9), optative (Fig. 10) and prohibitive (Fig. 11).

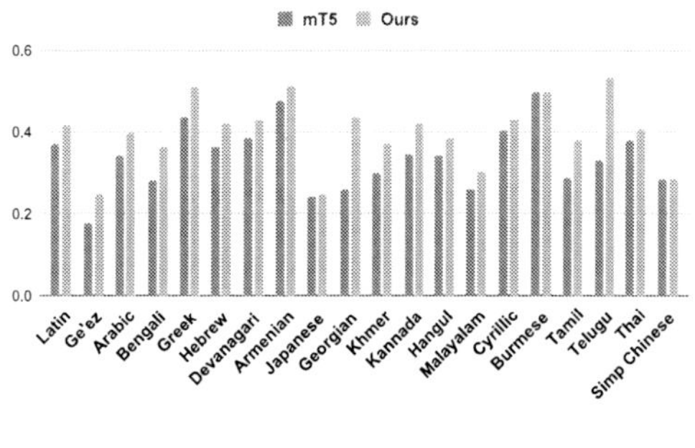

Figure 3: Exact match accuracy by language script.

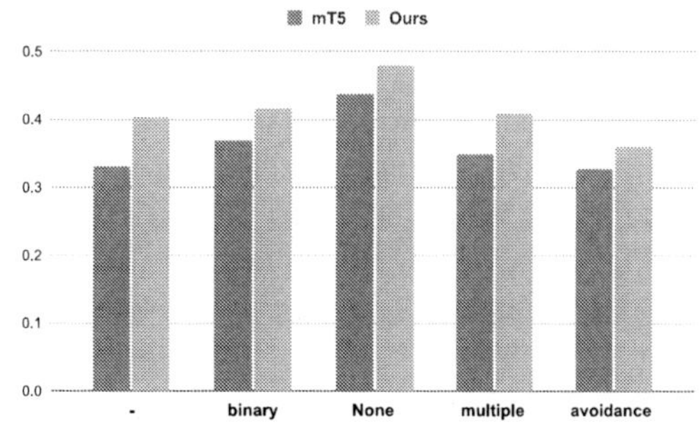

Figure 7: Exact match accuracy by language politeness.

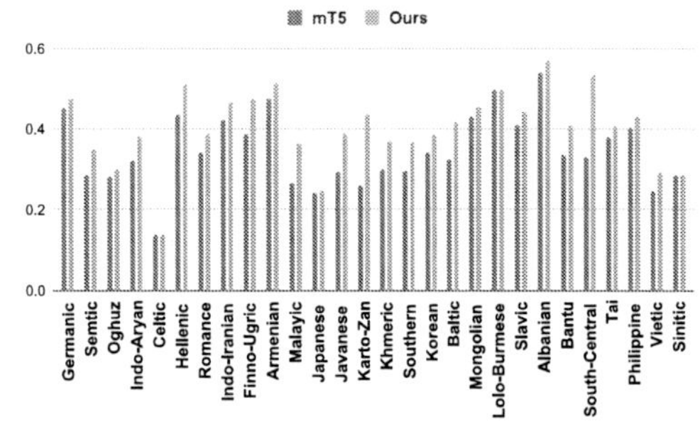

Figure 4: Exact match accuracy by language subdivision.

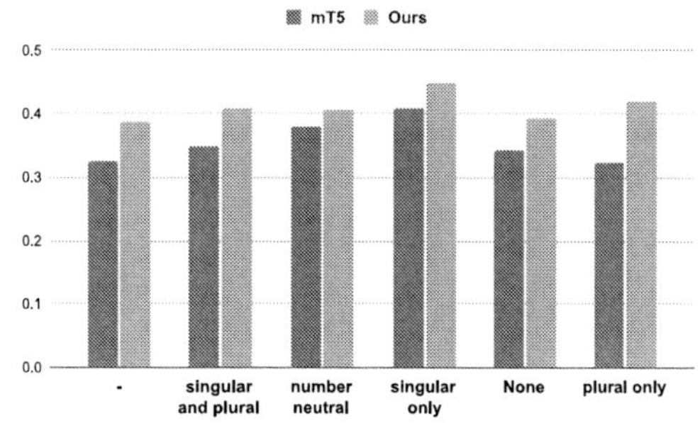

Figure 8: Exact match accuracy by language imperative morphology.

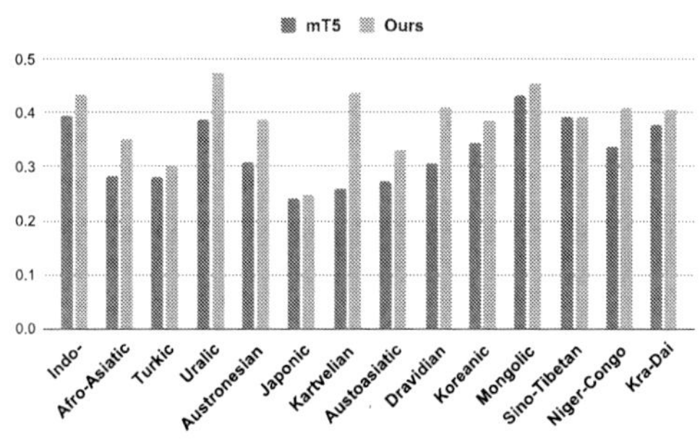

Figure 5: Exact match accuracy by language family.

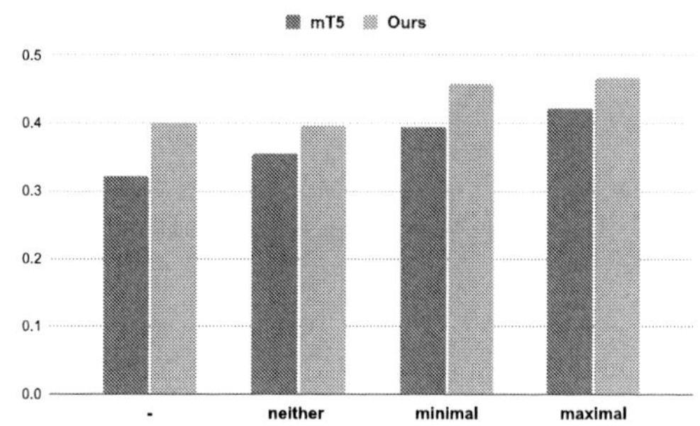

Figure 9: Exact match accuracy by language imperative hortative.

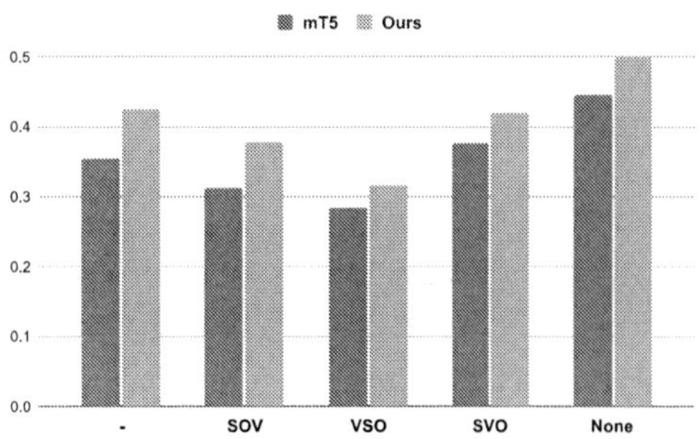

Figure 6: Exact match accuracy by language order.

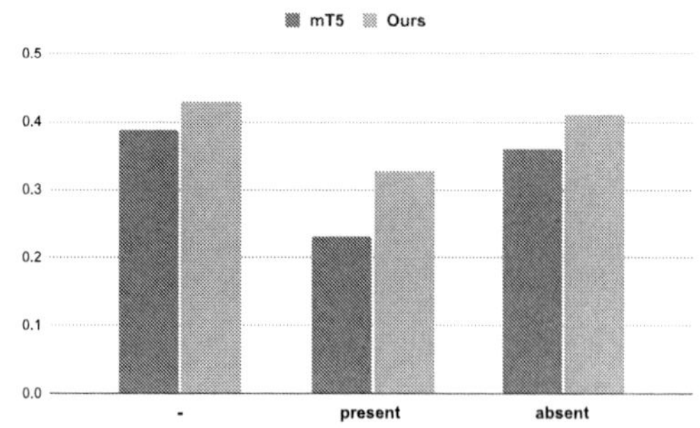

Figure 10: Exact match accuracy by language optative.

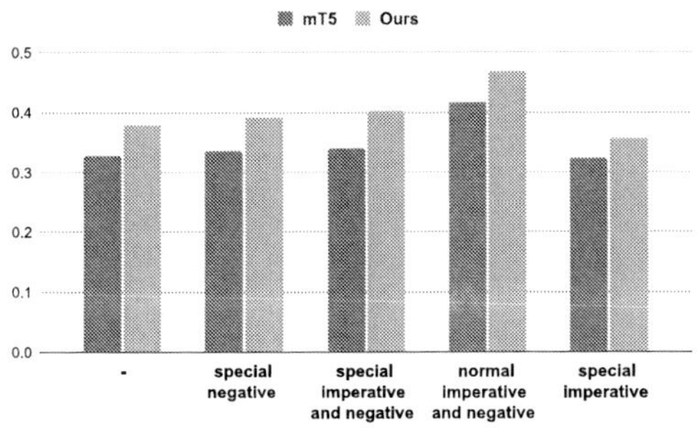

Figure 11: Exact match accuracy by language prohibitive.

C5L7: A Zero-Shot Algorithm for Intent and Slot Detection in Multilingual Task Oriented Languages

Jiun-Hao Jhan[1], Qingxiaoyang Zhu[2], Nehal Bengre[3], and Tapas Kanungo[3]

[1]Carnegie Mellon University Silicon Valley, CA, USA
[2]University of California Davis, CA, USA
[3]Samsung Research America, CA, USA
[1]*jiunhaoj@andrew.cmu.edu*
[2]*qinzhu@ucdavis.edu*
[3]*{n.bengre, tapas.k}@samsung.com*

Abstract

Voice assistants are becoming central to our lives. The convenience of using voice assistants to do simple tasks has created an industry for voice-enabled devices like TVs, thermostats, air conditioners, etc. It has also improved the quality of life of elders by making the world more accessible. Voice assistants engage in task-oriented dialogues using machine-learned language understanding models. However, training deep-learned models take a lot of training data, which is time-consuming and expensive. Furthermore, it is even more problematic if we want the voice assistant to understand hundreds of languages. In this paper, we present a zero-shot deep learning algorithm that uses only the English part of the Massive dataset and achieves a high level of accuracy across 51 languages. The algorithm uses delexicalized translation to generate a multilingual parallel corpus with intent and slot labels for data augmentation. The training data is further weighted to improve the accuracy of the worst-performing languages. We report on our experiments with code-switching, word order, multilingual ensemble methods and other techniques and their impact on overall accuracy.

1 Introduction

Task-oriented languages have become standard in voice-enabled devices and voice assistants. While there has been extensive research on task-oriented dialogue systems in limited domains, most of these systems are built in a limited set of languages due to a lack of labeled multilingual corpus. Amazon's MASSIVE dataset is a new resource for task-oriented language understanding that has 996K utterances annotated with intent and slot labels, along with their translations into 51 languages. The MASSIVE dataset is a unique resource for conducting multilingual language understanding research, and in particular building zero-shot learning algorithms where using only one language data, the trained system can perform language understanding tasks in the rest of the unseen languages. The importance of such training algorithms cannot be understated – labeled data is expensive and time-consuming to generate and hence any approach that reduces the cost and time to train such a multilingual system is desirable.

There are numerous hurdles in creating a zero-shot multilingual language understanding system. While machine translation systems can be used for translating utterances and creating a parallel corpus for training, aligning slot labels across languages can be challenging. In addition, if we expect the multilingual model representation to leverage information across languages, the input text representation needs to have the same tokenization process across languages. Furthermore, low-density languages are hard to get open-source resources for.

In this paper, we first review the related work in Section 2. Next, we address the issues listed above by introducing a novel delexicalized annotated utterance translation algorithm that is described in Section 3. To align code representations across languages, we randomly switched the language for a small percentage of the words. Finally, we explored the possible impact of using all the utterance translations instead of just one utterance in a specific language and were surprised by the accuracy boost. These and other experiments and analyses are described in Section 4.

2 Related Work

Transformer-based large multilingual masked language models, such as mBERT (Devlin et al., 2018), XLM (Lam- ple and Conneau, 2019) XLM-R (Conneau et al., 2019; Goyal et al., 2021), and mT5(Xue et al., 2020), have prevailed in cross-lingual language understanding. These models are pre-trained on a large multilingual text corpus to create a language representation that allows cross-lingual transfer on down-streaming tasks, such as cross-lingual document classification (Schwenk and Li, 2018; Pappas and Popescu-Belis, 2017),

Proceedings of the Massively Multilingual Natural Language Understanding Workshop (MMNLU-22), pages 62 - 68
December 7, 2022 ©2022 Association for Computational Linguistics

role labeling (Björkelund et al., 2009), question answering (Kwiatkowski et al., 2019; Chen et al., 2017; Lewis et al., 2019) and named entity recognition (Nothman et al., 2013; Al-Rfou et al., 2015). In the field of natural language understanding, Liu and Lane and Chen et al. trained for intent classification and slot-filling tasks jointly to learn the inherent correlation between the two tasks via multitask learning. Castellucci et al. further used a joint Bert-based model to detect intents and extract slots for the multi-lingual scenario including English and Italian languages. However, systematic work on massive languages datasets (51 locales including sufficient variance of language order types including subject-initial, verb-initial and no preferred word order) has not been paid enough attention until now due to the lack of labeled datasets.

With the availability of the MASSIVE dataset (FitzGerald et al., 2022) with annotation for slot-filling and intent classification, and virtual assistant evaluation metrics and scoring tools, we will able to push the state of the art of multilingual natural language processing for a task-oriented dialogue system (Razumovskaia et al., 2022; Tur et al., 2010). To tackle the difficulty/high cost of collecting low-resource language data previous work (Xu et al., 2020; Upadhyay et al., 2018; Schuster et al., 2018) explored the use of machine translation to get translated data (Wu et al., 2016) and utilize zero-shot learning (Palatucci et al., 2009) to transfer the understanding learned on one language to another language. However, the correspondence across all languages in terms of intent and slot alignment is insufficiently incorporated into the training and inference phases of the cross-lingual NLU model. In this paper, we explore how to represent connection among massive languages in the model. Besides, inspired by the common code-switch behavior and multilingual speakers and previous work on learning cross-lingual structure (Heredia and Altarriba, 2001; Wu et al., 2019; Auer, 2013), we further explore the use of code-switch and delexicalization as anchor points to bridge the transfer learning among languages.

3 Method

3.1 Data Augmentation

Generated Parallel Corpus To train a zero-shot learning model, using English data is not sufficient, as can be seen in the low baseline results in Table 1. To address the problem, we propose to utilize Google Translator to translate English data to the other 50 languages and create an annotated parallel corpus. While translating an utterance is simple, translating annotated utterances is difficult since the alignment of the slots like "time" and "date" is not always straightforward. Our solution is to delexicalize the slots in the given utterance (described in Section 3.2) and use the delexicalized utterance as input to Google Translator. Next, we create a lookup table to map the delexicalized slots and the slot values. We then translate the original slot values into the target language. Finally, we use the lookup table to substitute the translated slots values into the corresponding delexicalized tags in the translated utterance. This process results in a translated annotated utterance in the target language. Each annotated English utterance is thus translated into each of the 50 target languages while preserving the intent and slot annotations.

Method	Intent	Slot F1	Exact Match
Baseline	70.6 %	50.3 %	38.7 %
GPC	79.7 %	58.8 %	40.3 %
GPC+DE	81.1 %	58.84 %	40.3 %

Table 1: This table shows the comparison between using generated parallel corpus (GPC), delexicalization (DE) and not using delexicalization. Baseline results are for our implementation of Zero-shot Intent and Slot Prediction algorithm by FitzGerald et al. (2022). We see that augmenting the training data with the generated parallel corpus (GPC) gives us a significant boost to intent and slot accuracy. When we add delexicalized utterances in addition to GPC, (GPC + DE), we get a further boost to intent accuracy, but not much to slot accuracy.

Method	Full-Dataset		
	Intent	Slot F1	Exact Match
Baseline	85.10 %	73.60 %	63.70 %
BOS	85.72 %	75.01 %	65.12 %
BOS + LO	85.87 %	74.75 %	65.20 %

Table 2: Objective Functions Results. We evaluated three objective functions with the full dataset (instead of zero-shot learning). Baseline results are for our implementation of Intent and Slot Prediction algorithm by FitzGerald et al. (2022). BOS means the Bag of Slot and LO means the language word order prediction. These objective functions give slight improvement to slot accuracy.

Augmentation for Low-Performing Languages Low-performing languages decrease the total performance dramatically. Augmenting data for low-

Language Order	Order-Specific Models			All Language Model		
	Intent	Slot F1	Exact Match	Intent	Slot F1	Exact Match
ALL	-	-	-	85.66 %	75.12 %	65.35 %
SVO	86.23 %	74.54 %	64.84 %	86.18 %	74.65 %	65.00 %
SOV	75.69 %	63.67 %	50.53 %	85.11 %	74.50 %	64.60 %
VSO	66.55 %	64.69 %	43.20 %	84.00 %	72.43 %	62.14 %
Uncategorized	77.88 %	69.72 %	54.25 %	86.03 %	74.41 %	64.74 %
None	82.31 %	70.73 %	58.76 %	86.37 %	74.49 %	65.47 %

Table 3: Languages Word Order Results. Order-Specific Models mean the models are trained on a specific language word order class and evaluated in the same class. All Language Model is trained jointly with all languages and evaluated on a specific language word order. Using all languages improves accuracy.

performing languages is one possible approach to address this issue. We collect the lowest performing ten languages and reweight the data by 2x and 5x.

3.2 Code Switching

To align model representation across languages, researchers (Lee et al., 2019) have used the notion of "code-switching," where they randomly switch the language of a small percentage of the words in the training corpus. We used a similar approach in our model training process. We identified common stop words across languages and used their English translations for random code-switching. For non-space separated languages ("zh-CN", "zh-TW", "ja-JP"), we do code-switching with 8%, while the rest languages are with 16% of the words. Code-switching potentially creates anchor points (the common sequences in different languages) across multiple languages and assists transfer learning.

3.3 Delexicalized Training Data

Earlier, we used slot delexicalization to generate the parallel multilingual corpus for training data augmentation. In this section, we use delexicalization for a different purpose. We use slot delexicalization to learn slot *usage patterns*. We delexicalized the slots randomly. The various slot values are replaced by slot types. For example, the annotated utterance "Wake me up at [time : five am] [date : this Friday]" is delexicalized to "Wake me up at TIME_SLOT DATE_SLOT". We delexicalize utterances in each language to learn shared features in the multilingual dataset. We delexicalize the input utterance slots with a probability of $\epsilon = 0.1$ while training.

3.4 Objective Functions

We represented the problem as a multi-task recognition problem. The models were initialized with a pre-trained XLM-Roberta (XLM-R) (Conneau et al., 2020) language model and fine-tuned it on the MASSIVE dataset (mas). We then trained four different classification heads from scratch: intent and slots prediction, bag of slot labels, and language order prediction in parallel.

Intent and Slot Prediction Our model is aimed to do intent classification and slot-filling tasks in the zero-shot scenario. We used the training process described in mas. We use the English subset of the data and augment it with our (generated) annotated parallel corpus as described in Section **??**. For intent classification, the model predicts the intent by using the pooled output from the XLM-R encoder which is the sentence-level embedding vector. Then, the model predicts slot logits (as a sequence labeling task) using XLM-R encoder representations of each token in the utterance. Then the CrossEntropy loss function is used to compare the intent and slot logits with ground truth labels to get the intent and slot loss.

Bag of Slot Labels (BOS) Since each utterance has 51 translated versions, we leverage the constraint that all 51 utterances have the same intent and slot labels. We batched the English utterance and the corresponding utterances in other languages into one block. The meaning of the utterances in the unit is the same. The only difference is that they are written in different languages. We expect the predictions within a unit to be as similar as possible. Thus, in this block of parallel multilingual utterances, ideally, each of the utterances should predict the same slot labels. (Although the slot labels across languages may not be aligned at each token, the set of *B-SLOTNAME* and *I-SLOTNAME* slot tags (in the BIO format) in each utterance inside a batch is the same as others. We represent the bag of slot labels as a D_{slots} dimensional binary vector with each location indicating which slots labels are present in an utterance, where D_{slots} is the number of slot labels.) We collect 51 predictions as the output of intent classification and slot filling. Then we apply the CrossEntropy loss between the 51 intent predictions with ground truth.

Since the number of words in an utterance across the 51 languages and their word order might be different, computing loss per token does not work since the tokens are not aligned across languages. Thus, we get the mean of 51 languages' slot predictions and calculate the frequency of each slot type among these 51 utterances. Computing the CrossEntropy loss between the mean slot label predictions and the frequency might align the slot label predictions across the 51 predictions.

Language Word Order Prediction (LO) Word order is important in language. There are complicated rules for ordering words in different languages: two same utterances in different languages might generate large differences in the word's position in the sentence. Some languages start a sentence with the subject (S) following the verb (V) and the object (O). Others might start with the verb and end with the object. Therefore, we create another head to predict the language word order given an input utterance, training on the MASSIVE dataset. There are 5 kinds of word order in the MASSIVE dataset, SVO, SOV, VSO, none type, and uncategorized. We compute the CrossEntropy loss function between the order prediction and the ground truth. This loss function acts as one of the multitask among our objective functions.

4 Experiments and Results

4.1 Impact of Generated Parallel Corpus

The original baseline zero-shot algorithm described in mas uses only the English subset of the MASSIVE dataset and fine-tunes the multilingual XLMR model. We first explore the impact of augmenting the English subset of MASSIVE dataset with our generated (annotated) parallel corpus. In Table 1 we can see that our data augmentation increases the intent accuracy by 9.1% absolute and improves the average slot F1 score by 8.5%.

4.2 Augmenting Delexicalized Utterances

Next, in addition to augmenting the data with the generated parallel corpus, we added the delexicalized utterances. Table 1 shows that after applying the delexicalization technique, the intent accuracy increased by an absolute 2%. However, delexicalization barely improves the slot F1 score. The delexicalized data represents utterance templates, which the model learns, and perhaps helps with the intent accuracy. It is unclear why the slot accuracy was not impacted, perhaps a higher probability of delexicalization will help.

4.3 Objective Functions Comparison

In this experiment, we evaluate three objective functions by training on the full dataset from the MASSIVE (not zero-shot) training setup and testing on the corresponding test set, as shown in Table 2. The baseline results of the Intent and Slot prediction objective function are our implementation of (FitzGerald et al., 2022).

After including the Bag of Slot (BOS) objective function, the Slot F1 score increased by 2%. The main reason is that our model is capable of leveraging the shared information among 51 languages. However, adding the language word order prediction (LO) did not improve the performance. We found that the accuracy of language word order prediction is close to 100% and the loss is close to 0. The implication is that the XLMR model has learned to classify the language word order very well. However, the constraint of predicting language word order barely influenced the overall result.

4.4 Language Word Order Prediction Results

In this experiment, instead of training all languages jointly, we trained five different models corresponding to the language word order. In FitzGerald et al. (2022) the authors classified the language word order into five classes, SVO, SOV, VSO, Uncategorized, and None. According to results presented in Table 3, training a single SVO class model gets a similar performance as training on all languages jointly, while other classes get worse results. The main reason is that languages in the SVO class, like English, Spanish, etc., dominated the dataset. XLMR pretrained model is capable of understanding languages in the SVO class well. As for other language word orders, there might be a low-resource data problem while training the pretrained model that gives rise to a huge accuracy difference with respect to SVO class languages. In addition, training with all languages gives us better performance than training with only one language. The reason might be that training jointly makes the model leverage common characteristics amongst different language word orders.

Method	Full-Dataset				
	Training Method	Test Dataset	Intent	Slot F1	Exact Match
Amazon XLM-Base	full-training	MMNLU test	85.10 %	73.60 %	63.70 %
Amazon XLM-Base	zero-shot	MMNLU test	70.6 %	50.30 %	38.70 %
XLM-Base +BOS+DE	zero-shot	MMNLU-22 test	81.55 %	59.26 %	40.49 %
XLM-Base+GPC +BOS+DE+Ensemble	zero-shot	MMNLU-22 test	88.13 %	59.42 %	42.08 %

Table 4: Ensemble Result on MMNLU-22 Test Split. We evaluated our final model (the model with BOS and DE in training, Ensemble in post-processing) with MMNLU-22 test split, which is the test split of MMNLU-22 competition zero-shot track. The model was trained with the GPC dataset. The result of Amazon XLM-Base's model using full data training and the result of Amazon XLM-Base using zero-shot training on en-US are referred from the original MMNLU paper FitzGerald et al. (2022). BOS means the Bag of Slot, DE means delexicalized, and GPC means generated parallel corpus. The ensemble strategy gives significant improvement to intent accuracy on MMNLU-22 test set, making it even higher than Amazon's full-training dataset baseline results on MMNLU test set.

4.5 Post Processing with Ensemble Method

To leverage the characteristic of the parallel dataset, we experimented with an ensemble technique. Since for each utterance we have 50 translations with the same intent, we make each language vote for an intent and select the intent with the most votes as the final intent for all 51 languages. As a result (shown in Table 4), our model, including three objective functions and the voting technique, achieves 88.13%, 59.42%, and 42.08% for intent accuracy, slot F1 score, and exact match accuracy, respectively in MMNLU-22 test split[1]. In fact, intent accuracy achieves a significant boost with 6.61% in comparison to the result without ensemble strategy. We also see that the resulting intent accuracy is higher even than Amazon's baseline full-training data set. The slot F1 score, though significantly higher than Amazon's zero-shot baseline, is still much lower than the full-training data set results. This is probably due to using translations of English slot values to target languages. In our experiment, we used the translations from English to the target languages. However, to apply the voting technique in practice, we need to translate the utterance in the input language to all other target languages to elicit our model's multi-perspective on other languages and get a robust prediction through the ensemble. This work is currently in progress.

[1]The website of the competition with leaderboard: `https://eval.ai/web/challenges/challenge-page/1697/leaderboard/4061`

5 Discussion and Future Work

How good is the delexicalized slot translation? One approach to quantify this would be to generate an annotated translation from English to language i using Google translator and then translate it back to English and then compute a BLEU score.

Our zero-shot ensemble method using generated parallel corpus gives us better intent accuracy than the baseline full-set result in (FitzGerald et al., 2022). However, the slot accuracy is still much lower. One of the reasons could be that the slot values don't translate well to other languages. For example, a Christian name is not something that will be common in Chinese data. Using language-specific values probably will yield better results.

The ensemble method in a real-world setting requires us to translate utterance t in language i, to all other 50 languages. This requires to generate n^2 translation, which is expensive on Google Translate. For our experiment, we instead used the translations from English. One issue with this approach could be that English-to-target language translation might be of better quality than the translation of input language to a target language. Doing the full experiment will be conclusive. Another drawback of the ensemble approach is the need for n real-time translations and n parallel real-time runs. However, one way to reduce this complexity is to find a small subset of languages that we can use for voting purposes.

6 Conclusion

We presented a zero-shot, multilingual, joint intent-detection and slot-filling algorithm based on XLM-

R Transformer and Amazon's MASSIVE dataset. We showed that our delexicalized translation approach to generating a parallel corpus for data augmentation is a viable approach for training zero-shot algorithms. We showed that training using data from all language order types gives superior accuracy than using only a single language order type data in most cases – n MASSIVE data, the SVO category performed equally well when using just the SVO subset. Furthermore, our experiments showed that using an ensemble approach with translations of the input utterance can lead to a significant gain in accuracy.

Acknowledgements

We would like to thank the anonymous reviewers for their insightful comments on an earlier draft of this paper and Steve Walsh and Yuri Lozhnevsky for their assistance.

References

Rami Al-Rfou, Vivek Kulkarni, Bryan Perozzi, and Steven Skiena. 2015. Polyglot-ner: Massive multilingual named entity recognition. In *Proceedings of the 2015 SIAM International Conference on Data Mining*, pages 586–594. SIAM.

Peter Auer. 2013. *Code-switching in conversation: Language, interaction and identity*. Routledge.

Anders Björkelund, Love Hafdell, and Pierre Nugues. 2009. Multilingual semantic role labeling. In *Proceedings of the Thirteenth Conference on Computational Natural Language Learning (CoNLL 2009): Shared Task*, pages 43–48.

Giuseppe Castellucci, Valentina Bellomaria, Andrea Favalli, and Raniero Romagnoli. 2019. Multi-lingual intent detection and slot filling in a joint bert-based model. *arXiv preprint arXiv:1907.02884*.

Danqi Chen, Adam Fisch, Jason Weston, and Antoine Bordes. 2017. Reading wikipedia to answer open-domain questions. *arXiv preprint arXiv:1704.00051*.

Qian Chen, Zhu Zhuo, and Wen Wang. 2019. Bert for joint intent classification and slot filling. *arXiv preprint arXiv:1902.10909*.

Alexis Conneau, Kartikay Khandelwal, Naman Goyal, Vishrav Chaudhary, Guillaume Wenzek, Francisco Guzmán, Edouard Grave, Myle Ott, Luke Zettlemoyer, and Veselin Stoyanov. 2019. Unsupervised cross-lingual representation learning at scale. *arXiv preprint arXiv:1911.02116*.

Alexis Conneau, Kartikay Khandelwal, Naman Goyal, Vishrav Chaudhary, Guillaume Wenzek, Francisco Guzmán, Edouard Grave, Myle Ott, Luke Zettlemoyer, and Veselin Stoyanov. 2020. Unsupervised cross-lingual representation learning at scale. In *Proceedings of the 58th Annual Meeting of the Association for Computational Linguistics*, pages 8440–8451, Online. Association for Computational Linguistics.

Jacob Devlin, Ming-Wei Chang, Kenton Lee, and Kristina Toutanova. 2018. Bert: Pre-training of deep bidirectional transformers for language understanding. *arXiv preprint arXiv:1810.04805*.

Jack FitzGerald, Christopher Hench, Charith Peris, Scott Mackie, Kay Rottmann, Ana Sanchez, Aaron Nash, Liam Urbach, Vishesh Kakarala, Richa Singh, et al. 2022. Massive: A 1m-example multilingual natural language understanding dataset with 51 typologically-diverse languages. *arXiv preprint arXiv:2204.08582*.

Naman Goyal, Jingfei Du, Myle Ott, Giri Anantharaman, and Alexis Conneau. 2021. Larger-scale transformers for multilingual masked language modeling. *arXiv preprint arXiv:2105.00572*.

Roberto R Heredia and Jeanette Altarriba. 2001. Bilingual language mixing: Why do bilinguals code-switch? *Current Directions in Psychological Science*, 10(5):164–168.

Tom Kwiatkowski, Jennimaria Palomaki, Olivia Redfield, Michael Collins, Ankur Parikh, Chris Alberti, Danielle Epstein, Illia Polosukhin, Jacob Devlin, Kenton Lee, et al. 2019. Natural questions: a benchmark for question answering research. *Transactions of the Association for Computational Linguistics*, 7:453–466.

Grandee Lee, Xianghu Yue, and Haizhou Li. 2019. Linguistically motivated parallel data augmentation for code-switch language modeling. In *Interspeech*, pages 3730–3734.

Patrick Lewis, Barlas Oğuz, Ruty Rinott, Sebastian Riedel, and Holger Schwenk. 2019. Mlqa: Evaluating cross-lingual extractive question answering. *arXiv preprint arXiv:1910.07475*.

Bing Liu and Ian Lane. 2016. Attention-based recurrent neural network models for joint intent detection and slot filling. *arXiv preprint arXiv:1609.01454*.

Joel Nothman, Nicky Ringland, Will Radford, Tara Murphy, and James R Curran. 2013. Learning multilingual named entity recognition from wikipedia. *Artificial Intelligence*, 194:151–175.

Mark Palatucci, Dean Pomerleau, Geoffrey E Hinton, and Tom M Mitchell. 2009. Zero-shot learning with semantic output codes. *Advances in neural information processing systems*, 22.

Nikolaos Pappas and Andrei Popescu-Belis. 2017. Multilingual hierarchical attention networks for document classification. *arXiv preprint arXiv:1707.00896*.

Evgeniia Razumovskaia, Goran Glavas, Olga Majewska, Edoardo M Ponti, Anna Korhonen, and Ivan Vulic. 2022. Crossing the conversational chasm: A primer on natural language processing for multilingual task-oriented dialogue systems. *Journal of Artificial Intelligence Research*, 74:1351–1402.

Sebastian Schuster, Sonal Gupta, Rushin Shah, and Mike Lewis. 2018. Cross-lingual transfer learning for multilingual task oriented dialog. *arXiv preprint arXiv:1810.13327*.

Holger Schwenk and Xian Li. 2018. A corpus for multilingual document classification in eight languages. *arXiv preprint arXiv:1805.09821*.

Gokhan Tur, Dilek Hakkani-Tür, and Larry Heck. 2010. What is left to be understood in atis? In *2010 IEEE Spoken Language Technology Workshop*, pages 19–24. IEEE.

Shyam Upadhyay, Manaal Faruqui, Gokhan Tür, Hakkani-Tür Dilek, and Larry Heck. 2018. (almost) zero-shot cross-lingual spoken language understanding. In *2018 IEEE International Conference on Acoustics, Speech and Signal Processing (ICASSP)*, pages 6034–6038. IEEE.

Shijie Wu, Alexis Conneau, Haoran Li, Luke Zettlemoyer, and Veselin Stoyanov. 2019. Emerging cross-lingual structure in pretrained language models. *arXiv preprint arXiv:1911.01464*.

Yonghui Wu, Mike Schuster, Zhifeng Chen, Quoc V Le, Mohammad Norouzi, Wolfgang Macherey, Maxim Krikun, Yuan Cao, Qin Gao, Klaus Macherey, et al. 2016. Google's neural machine translation system: Bridging the gap between human and machine translation. *arXiv preprint arXiv:1609.08144*.

Weijia Xu, Batool Haider, and Saab Mansour. 2020. End-to-end slot alignment and recognition for cross-lingual nlu. *arXiv preprint arXiv:2004.14353*.

Linting Xue, Noah Constant, Adam Roberts, Mihir Kale, Rami Al-Rfou, Aditya Siddhant, Aditya Barua, and Colin Raffel. 2020. mt5: A massively multilingual pre-trained text-to-text transformer. *arXiv preprint arXiv:2010.11934*.

Machine Translation for Multilingual Intent Detection and Slots Filling

Maxime De Bruyn, Ehsan Lotfi, Jeska Buhmann, Walter Daelemans
CLiPS Research Center
University of Antwerp, Belgium
maxime.debruyn@uantwerpen.be

Abstract

We expect to interact with home assistants irrespective of our language. However, scaling the Natural Language Understanding pipeline to multiple languages while keeping the same level of accuracy remains a challenge. In this work, we leverage the inherent multilingual aspect of translation models for the task of multilingual intent classification and slot filling. Our experiments reveal that they work equally well with general-purpose multilingual text-to-text models. Furthermore, their accuracy can be further improved by artificially increasing the size of the training set. Unfortunately, increasing the training set also increases the overlap with the test set, leading to overestimating their true capabilities. As a result, we propose two new evaluation methods capable of accounting for an overlap between the training and test set.

1 Introduction

Home assistants are omnipresent in everyday life. We expect to have an assistant at our disposal at any time using our phone, watch, or car — irrespective of our language.

Scaling home assistants to multiple languages brings additional challenges to NLU and ASR components. There are two options: a single model per language or a shared model for all languages. A single model per language works well for resource-rich languages such as English. However, lower resource languages benefit from the cross-lingual knowledge transfer of a single model dealing with all languages (Conneau et al., 2020). This trade-off applies to any multilingual system (Zhang et al., 2022; De Bruyn et al., 2021).

While multilingual intent classification and slot filling datasets exist, their language coverage is limited, except for MASSIVE (FitzGerald et al., 2022), a new dataset focused on multilingual intent detection and slot filling. The authors translated and localized an English-only dataset in 50 topologically diverse languages. MASSIVE provides a good base to scale existing intent detection and slot filling methods to multiple languages.

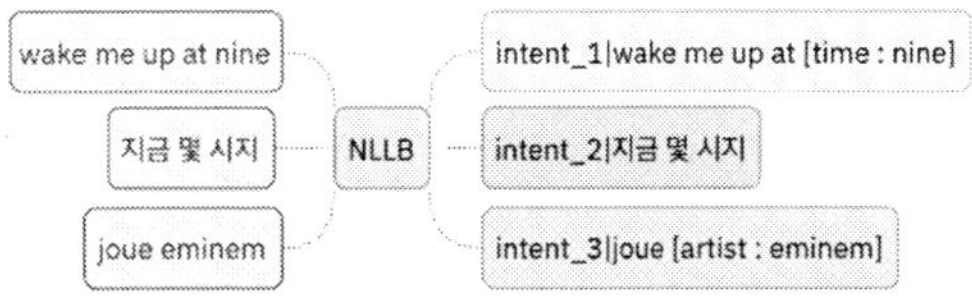

Figure 1: Illustration of our method. We repurpose a translation model for the task of multilingual intent classification and slot filling. We translate from utterances into annotated utterances.

The traditional way to tackle multilingual intents detection and slot filling is to use multilingual models such as XLM-R (Conneau et al., 2020), or mT5 (Xue et al., 2021). These models are similar to their monolingual counterparts (Liu et al., 2019; Raffel et al., 2020) except for the multilingual data used to train them.[1] This approach has been shown to work in multiple studies (FitzGerald et al., 2022; Li et al., 2021). However, MASSIVE has an additional overlooked aspect: utterances are direct translations of one another.

In this work, we approach the task of intent classification and slot filling as a translation task: we translate the original utterance into the annotated utterance. For example, we translate the utterance `what is the temperature in new york?` into the annotated utterance `weather_query|what is the [weather_descriptor : temperature] in [place_name : new york]`.[2]

The typical use of translation models for intent detection and slot filling is to augment the size of an existing dataset (Zheng et al., 2021; Nicosia et al., 2021). However, we believe the inherent multilingual capabilities of these models make them excellent candidates for multilingual intent detec-

[1] They also have larger vocabularies and may have special training tricks for cross-lingual training.

[2] We prepend the slot annotated utterance with the intent.

Proceedings of the Massively Multilingual Natural Language Understanding Workshop (MMNLU-22), pages 69 - 82
December 7, 2022 ©2022 Association for Computational Linguistics

tion and slot filling.

To this end, we leverage the recently released translation model *No Language Left Behind* (NLLB) (NLLB Team et al., 2022) capable of translating between 202 pairs of languages simultaneously using a shared encoder-decoder. We anticipate that the wide range of languages covered by the model will help us deal with lower resources languages present in the MASSIVE dataset.

Better modeling is only half the story. Using more data also helps improve performance. For example, although the MASSIVE dataset displays a large training set of more than 500K training examples, the seed data is only around 10K training examples. Therefore, we used GPT-3 (Brown et al., 2020) to generate additional training data using a dual-model approach. We also leveraged a dataset close to the seed dataset of MASSIVE. As a result, after translating our new training examples to the 50 remaining languages, our training set contains more than 2M training examples — 4x the size of the original training set.

Our experiments reveal that translation models such as NLLB are a good fit for intent classification and slot filling. However, their performance sharply drops in languages that do not use spaces because of tokenization issues.

Unfortunately, the additional training data significantly overlaps with the MASSIVE test set. As a result, we propose two methods capable of dealing with overlaps: weighted exact match and logistic regression.

We conclude this introduction by summarizing our contributions:

- We showed that a translation model such as NLLB can complete the task of intent classification and slot filling
- We demonstrated a method to improve the training data with GPT-3
- We proposed two new evaluation methods taking the training/test set overlap into account

We release our model[3], utterance translation model[4], and generated data[5] on the HuggingFace hub.

[3]maximedb/nllb_massive

[4]maximedb/massive_en_translation

[5]maximedb/massive_generated

2 Related Work

The problem of multilingual intent detection and slot filling is not new. (Razumovskaia et al., 2022) provides an excellent introduction to the subject. We divide our related work section into three parts. We start by reviewing the general problem of task-oriented semantic parsing (i.e., intent detection and slot filling). Next, we review the models commonly used, and lastly, we review the available multilingual datasets.

2.1 Task Oriented Semantic Parsing

Natural Language Understanding (NLU) systems aim to classify an utterance into a predefined set of intents and label the sequence with a predefined ontology of slots (McTear, 2020). Since the release of the ATIS dataset (Price, 1990), this problem has been studied in numerous previous work (Mesnil et al., 2013; Liu and Lane, 2016; Zhu and Yu, 2017). However, it has recently been shown that the flat structure of sequence labeling falls short when a user issues sub-queries, or compositional queries, e.g., `set up a reminder to message mike tonight`[6] Gupta et al. (2018) solves that problem by using hierarchical representations instead.

2.2 Translation Models

Previous work tackling multilingual intent detection and slot filling uses multilingual versions of well-known Transformers such XLM-Roberta (Conneau et al., 2020), mT5 (Xue et al., 2021), or mBART (Liu et al., 2020). We diverge from existing research and use machine translation models instead. (Fan et al., 2021) released M2M100, a model capable of translating between pairs of 100 languages using a single shared encoder-decoder model. Instead of mainly going from and to English, the authors use a dataset that covers thousands of language pairs. M2M100 was later improved by the release of No Language Left Behind (NLLB) (NLLB Team et al., 2022), which follows the same architecture as M2M100 but covers 202 languages.

2.3 Cross-Lingual Task Oriented Semantic Parsing

Although the initial dataset for intent classification and slot filling targeted English, the number of non-English datasets is growing rapidly. Non-English

[6]Two intents compose that query: create a reminder and send a message to mike.

datasets fall into two broad categories: non-English monolingual datasets (Meurs et al., 2008; Castellucci et al., 2019; Bellomaria et al., 2019; Zhang et al., 2017; Gong et al., 2019; He et al., 2013; Dao et al., 2021) and multilingual datasets. As we aim to study models capable of handling multiple languages simultaneously, we focus on the latter kind of datasets. We will now cover the existing multilingual datasets in greater detail. Upadhyay et al. (2018) translated an existing English dataset (Price, 1990) into Turkish and Hindi, while Susanto and Lu (2017) translated the same dataset in Vietnamese and Chinese. Schuster et al. (2019) released a multilingual dataset for task-oriented dialogues in English, Spanish, and Thai across three domains. (Li et al., 2021) provides MTOP a new aligned task-oriented dataset in six languages. MASSIVE (FitzGerald et al., 2022) is the largest available dataset, covering 51 languages.

3 Data

There exist multiple alternative datasets to study multilingual intent detection and slot filling. However, in this work, we use the largest one available: the MASSIVE dataset.

3.1 MASSIVE

MASSIVE (FitzGerald et al., 2022) is a dataset assembled by translating and localizing an existing English-only dataset in 50 topologically different languages.

English Seed MASSIVE is a translation of the English-centric SLURP dataset (Bastianelli et al., 2020). SLURP is a dataset of non-compositional queries directed at a home assistant. It covers 18 domains, 60 intents, and 55 slots.

Languages The authors of MASSIVE hired professional translators to translate the SLURP dataset into 50 topologically diverse languages from 29 genera. Furthermore, to complicate the task, the translators sometimes localized the queries instead of simply translating them.

3.2 English Data Augmentation

As the seed data of MASSIVE is limited in scale (10K training examples), we used two methods to increase the training set artificially.

3.2.1 Generated Data

Generator We first fine-tune a GPT-3 (Brown et al., 2020) curie (13B) model on the task of generating an English utterance conditional on the given intent. For example, we train the model to generate `wake me up at nine am` given the prompt `alarm_set`.

Parser Next, we fine-tune a second GPT-3 curie model on intent detection and slot filling tasks. Given an utterance, the model must generate the concatenation of the intent and the annotated utterance. For example, given the prompt `what is the temperature in new york?` must generate `weather_query|what is the [weather_descriptor : temperature] in [place_name : new york]`.

Dataset We generate 30,000 utterances, equally distributed amongst the 60 intents. After removing duplicates and examples where the two models do not agree on the intent, we arrive at a final dataset of 22,276 annotated English utterances.

Intent & Slots Distribution Although we generated an equal amount of utterances per intent, removing duplicates skewed the distribution. However, comparing the entropy of both distributions with MASSIVE reveals that our generated dataset is more equally spread amongst the intents but less equally distributed relative to the slots.[7] See Annex A for a detailed analysis and comparison with the MASSIVE dataset.

3.2.2 Synthetic Data

The SLURP dataset provides a synthetic dataset.[8] It is not part of the official training set, but as it shares the same ontology as MASSIVE, it provides an excellent extension to our training set. We compare the intent and slot distribution with MASSIVE in Annex A.

3.3 Non-English Data Augmentation

We explained in Section 3.2 our method to artificially increase the size of the (English) training set. This section reviews our method to scale this silver training set to the 50 remaining languages in the MASSIVE dataset.

Using commercial translation systems was not an option as this requires aligning the slots in the translated utterances — a complicated task. Instead, we fine-tune a translation model, NLLB (3B), on the task of translating *annotated* utterances directly.

[7]Our generated dataset has an intent distribution entropy of 4.02 and a slot distribution entropy of 3.10 compared to 3.75 and 3.21 for MASSIVE.

[8]https://github.com/pswietojanski/slurp/tree/master/dataset/slurp

Using this method, we translate annotated utterances and reconstruct the utterances by removing the slot annotations from the text. Our translation model is available on the HuggingFace Hub.[9]

4 Model

This work uses a machine translation model for intent detection and slot filling. No Language Left Behind (NLLB) (NLLB Team et al., 2022) is a model specifically targeted at translating between 202 languages using a single encoder-decoder model based on the M2M100 architecture (Fan et al., 2021). It can translate text in 40,602 different directions.

Data NLLB uses FLORES-200 as training data, an extension of FLORES-100 (Goyal et al., 2022). The authors of FLORES-200 used LASER3 (Heffernan et al., 2022) to mine parallel data from the web, resulting in 1.1 billion sentence pairs.

Tokenization NLLB uses a sentencepiece tokenizer (Kudo and Richardson, 2018) with a vocabulary size of 256,000. To ensure low-resource languages are well-represented in the vocabulary, the authors downsample high-resource and upsample low-resource languages.

Architecture NLLB's architecture is based on the Transformer encoder-decoder (Vaswani et al., 2017). NLLB is trained on several translation directions at once, utilizing the same shared model capacity. This architecture can lead to beneficial cross-lingual transfer between related languages at the risk of increasing interference between unrelated languages. The authors also present a Sparsely Gated Mixture of Experts (MoE) (Almahairi et al., 2016; Bengio et al., 2013). However, we did not experiment with this variant.

Distillation The authors distilled a 54 billion parameter model using MoE into smaller dense models of 1.3 billion and 615 million parameters using online distillation (Hinton et al., 2015). The student model is trained on the training data but with an additional objective: to minimize the cross-entropy to the word-level distribution of the teacher model. We use the distilled 615M parameter model as the base model for intent classification and slot filling.

[9]For anonymity reasons, we will release the URL upon acceptance of this paper.

5 Experiments

This section describes our experiments in applying NLLB to the task of intent classification and slot filling. NLLB is a translation model. While we could repurpose NLLB to the task of intent classification and slot filling directly, we choose to first pre-train it on a translation task.

5.1 Pre-training

As NLLB is, at its core, a translation model, we start by teaching it to translate between the aligned pairs of the MASSIVE dataset. Instead of translating between the utterances of two languages, we translate between the utterance and the annotated utterance. For example, the model must translate "tell me the time in moscow," to the French annotated utterance `datetime_query|donne moi l'heure à [place_name: moscou]`. We take special care in avoiding localized utterances, as this would confuse the model. For example, we avoid predicting `datetime_query|donne moi l'heure à` ~~moscou~~ `bordeaux`.

5.2 Fine-tuning

In a second step, we fine-tune the model on the task of translating between the utterance and the annotated utterance in the same language. For example, we translate the utterance `what is the temperature in new york?` into the annotated utterance `weather_query|what is the [weather_descriptor : temperature] in [place_name : new york]`.

5.3 Technical Details

We use the NLLB-200 (600M) model for all experiments.[10] We wrap each encoder input according to the following formula: `<s>...</><language_code>`. We prepend each decoder input with the target language code. We train for 50,000 steps during pre-training and fine-tuning with a learning rate of 1e−4 and 1e−5, respectively. We use Pytorch (Paszke et al., 2019), the HuggingFace Trainer (Wolf et al., 2020) and DeepSpeed (Rajbhandari et al., 2020).

6 Results

This section presents a high-level analysis of our results. Table 1 compares our results against the baselines provided by the authors of MASSIVE.

[10]facebook/nllb-200-distilled-600M

Model	Training Set	Intent Acc (%)			Slot F1 (%)			Exact Match (%)		
		High	Low	Avg	High	Low	Avg	High	Low	Avg
XLM-R	M	88.3	77.2	85.1	83.5	63.3	73.6	70.1	55.8	63.7
mT5 Enc.	M	89.0	79.1	86.1	85.7	64.5	75.4	72.3	57.8	65.9
mT5	M	87.9	79.0	85.3	**86.8**	67.6	76.8	73.4	58.3	66.6
NLLB	M+G	89.3	79.2	87.3	85.9	66.3	77.0	74.1	57.8	68.3
NLLB	M+G+S	**94.5**	**84.5**	**93.4**	82.9	**69.6**	**82.9**	**89.2**	**65.0**	**78.5**

Table 1: Modelling results on the MASSIVE test set. NLLB trained on the MASSIVE training set (M), our generated dataset (G) and the synthetic training set from SLURP (S) achieve the highest scores. However, as we show in a later section, this outperformance is due to a large overlap with the MASSIVE test set.

Our experiments reveal that NLLB performs similarly to mT5 on intent detection and slot filling tasks. Furthermore, our two data augmentation strategies improve the results on the MASSIVE test set. First, training with our generated training set improves the locale average exact match from 66.6 to 68.3. Second, training with the generated and synthetic data boosts the exact match as it improves from 68.3 to 78.5. As we show in the next section, this performance boost is mainly due to a large overlap between the training and test set.

Training Set	en-US (%)	All Locales (%)
MASSIVE	0.7	5.9
Generated	5.6	6.4
Synthetic	49.0	12.8

Table 2: Exact duplicate analysis. Percentage of examples in the MASSIVE test set, which are also present in the training set of MASSIVE, our generated training set, and the synthetic training set. Translation reduces the overlap of the synthetic dataset compared to the English-only figures. However, it is the opposite for the MASSIVE test set, where the overlap is higher for all locales compared to English only.

7 Training & Test Set Overlap

This section analyses the similarity between the training sets and the MASSIVE. Next, we look for evaluation methods capable of correcting for the overlap between the training and test set.

Exact Duplicates An analysis of the data reveals problematic overlaps between the training sets and the MASSIVE test set. However, this overlap is unequal across the training sets and languages. Table 2 shows the percentage of examples in the MASSIVE test set, which are also present in our three training sets. The English subset of the MASSIVE test set overlaps highly with the synthetic training set described in Section 3.2.2. Localization and translation somewhat reduce the exact match overlap when looking at all languages, although it remains high. The MASSIVE and generated training sets also have a non-zero overlap with the MASSIVE test set.

Close Duplicates Some examples may not be exact duplicates but close duplicates. For example, `call the dentist` and `olly please call the dentist now`. We use character n-grams to measure the similarity between two utterances as similarity metric between two utterances. We search for the most similar training example for each example in the test and record their n-gram similarity.[11] Figure 2 shows the distribution of maximum similarity between the test set and our three training sets for the English subset and across all locales. It is clear from Figure 2 that the English synthetic dataset overlaps significantly with the English MASSIVE test set. However, as for the exact duplicates, the translation and localization process reduces this overlap but does not eliminate it.

A naive solution would be to remove training examples that overlap with the test set. However, how does one decide what is a close duplicate? Furthermore, as the training set grows, some overlap with the test is inevitable. We argue that the problem is not the training data but the evaluation metric. We need an evaluation metric capable of controlling for the overlap between the test and training sets.

7.1 Logistic Regression

Instead of looking at the simple exact match accuracy, we want to express the exact match accuracy as a function of the test/train similarity. One potential solution is to use logistic regression with similarity as the independent variable and exact match as the dependent variable.

[11]We do this search on a per-language basis.

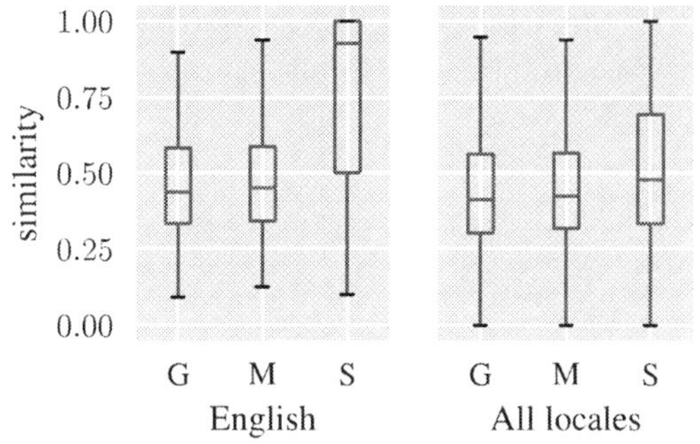

Figure 2: Box plot of the maximum similarity between examples in the MASSIVE test set with the training set of MASSIVE (M), Generated (G) and Synthetic (S), for the English part and the entire dataset (all locales). The English synthetic (S) training set overlaps highly with the MASSIVE test set. Translation and localization reduces this overlap in the all-locales dataset.

Training S.	β_0	β_1	R^2
M+G	-0.96±0.03	3.31±0.06	0.07
M+G+S	-0.69±0.03	3.14±0.06	0.08

Table 3: We report the logistic regression results for two NLLB models fine-tuned on the training set of MASSIVE (M), generated (G), and synthetic (S). We report the point estimate and the 95% confidence interval for each parameter. After correcting for any overlap between the training and test set, the second is statistically better than the first.

$$p(x) = \frac{1}{1 + e^{-(\beta_0 + \beta_1 x)}} \tag{1}$$

Where $p(x)$ represents the probability of an exact match, β_0 represents the intercept and β_1 the slope. Using this method, we can compare both models at the same level of similarity.

Results Table 3 presents a summary of the logistic regression results. We report the point estimate and confidence interval for both β_0, β_1 and the pseudo R^2 given by `statsmodels` (Seabold and Perktold, 2010). Using Equation 1, we can estimate the performance of both models at multiple levels of similarity, as shown in Figure 3.

According to Table 3 and Figure 3, the model trained on the three training datasets is better than the one trained only on two — taking the overlap into account. However, these numbers also indicate that both models struggle with utterances dissimilar to the training set. Moreover, they achieve an exact match accuracy lower than random chance on dissimilar utterances — casting doubt on their abilities to generalize to unseen utterances.

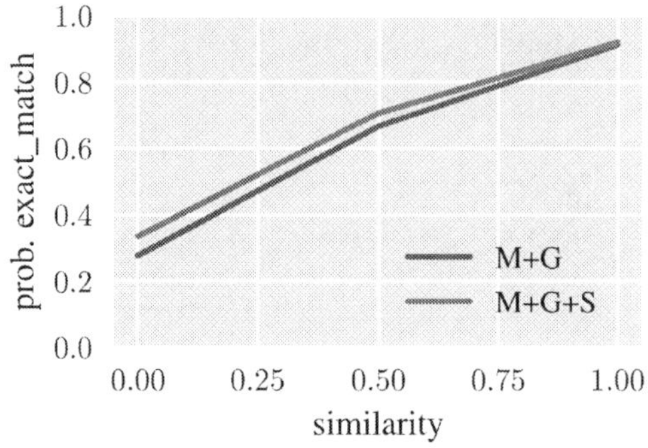

Figure 3: Exact match probability at three levels of similarity: 0, 0.5, and 1.0. We used Equation 1 with the estimated parameters from Table 3. Model two is better than model one on dissimilar utterances. However, the difference diminishes when the similarity increases.

Training Set	Weighted Average (%)
M+G	59.2
M+G+S	67.2

Table 4: We report the weighted average results for two NLLB models fine-tuned on the training set of MASSIVE (M), generated (G), and synthetic (S). The second model is better than the first even after correcting for its high overlap with the training set.

7.2 Weighted Average

Another possibility is to give less importance to test examples similar to the training set.

$$\sum_{i=1}^{n} \frac{w_i * exact\_match_i}{\sum_{i=1}^{n} w_i} \tag{2}$$

where $w_i = 1 - sim_i$.

Results Table 4 displays the results according to the weighted average metric. According to this metric, the second model outperforms the first one. This metric is easy to understand. However, it does not tell us anything about the performance of dissimilar queries.

7.3 Summary

According to our overlap-aware evaluation metrics, the model trained on the synthetic datasets is the most performant, even after correcting for its high overlap with the test.

Language	Intercept	Num. Token Split	R-Squared
ja-JP	0.85*	-0.16*	0.013
zh-CN	0.58*	-0.15*	0.006
zh-TW	0.11	-0.03	0.000

Table 5: Logistic regression of exact match accuracy explained by the number of split token. The number of split token negatively influence the capability of the token to correctly parse the slots for ja-JA and zh-CN. The coefficient are not significantly different than zero for zh-TW. Starred numbers (*) are statistically different than zero with a p-value of 0.05

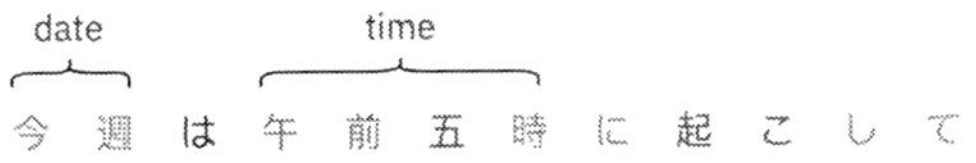

Figure 4: Our method does not scale well to non-space delimited languages. For example, in the utterance above, the time slot ends in the middle of a token. To correctly parse the utterance, the model must replace token 20202 (時に) by tokens 249229 (時) and 5954 (に).

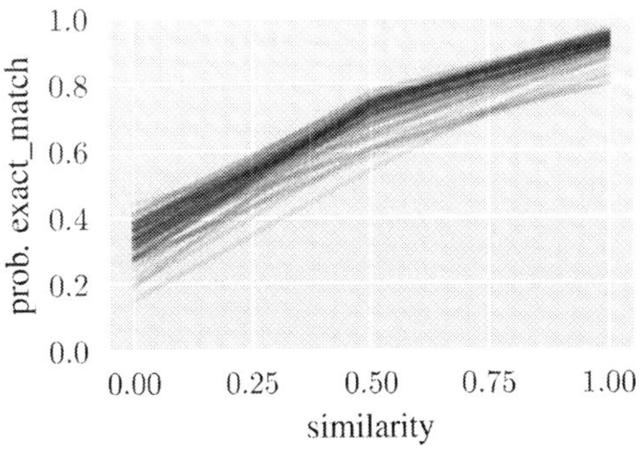

Figure 5: Exact match probability at three levels of similarity: 0, 0.5, and 1.0. We used Equation 1 with the estimated parameters from Table 3. Model two is better than model one on dissimilar utterances. However, the difference diminishes when the similarity increases.

8 Error Analysis

8.1 Tokenization

Our formatting of input and output consists of surrounding slots with brackets along with the slot name (e.g., `[place_name : new york]`. This method implies that slots' boundaries align with tokenization. Otherwise, the model cannot correctly place the opening or closing bracket — unless it uses a different token than the ones in the source utterance. See Figure for an example.

We identified three languages for which this problem occurs: ja-JP in 66% of the test set, zh-CN in 66% of the test set, and zh-TW in 69% of the test set. These are three languages that do not use spaces between words.

Similar to Section 7.1, we ran a logistic regression to explain the exact match performance by the number of split tokens. Table 5 shows the results. We identified a statistically significant relationship between the number of split tokens and the exact match performance for ja-JP and zh-CN. The performance of zh-TW is low regardless of the number of split tokens.

8.2 Generalization

Section 7.1 demonstrated that models struggle to generalize to utterances dissimilar to the training set. In this section, we decompose this conclusion by languages. Figure 5 decomposes Figure 3 by languages. It shows the probability of an exact match on the test set by increasing levels of similarity to the training set. Figure 5 shows a wide distribution of probabilities for low similarity utterances (6% standard deviation), while the distribution for highly similar utterances is more concentrated (3% standard deviation). Some languages do better than others. For example, km-KH achieves an exact match probability of 44% at a similarity of 0.0 while vi-VN only achieves a an exatch match probability of 15%. We list the full details of Figure 5 in Appendix B.

9 Future Work

In this work, we estimated the similarity between two utterances using character n-grams. However, while this captures lexically similar utterances, it fails to capture utterances semantically similar but lexically different. For example, these two utterances are highly similar, although they only share a single common token: `what time is it?` and `tell me the time`. Future work can tackle this by using multilingual sentence encoders such as LASER3 (Heffernan et al., 2022), Multilingual Universal Sentence Encoder (Yang et al., 2020), or multilingual models on Sentence Transformers

(Reimers and Gurevych, 2020).

This work did not explicitly address cross-lingual training and instead relied on the cross-lingual pre-training of the translation model. Future work could combine a translation model with cross-lingual training methods such as xTune (Zheng et al., 2021), or X-Mixup (Yang et al., 2022).

Section 8.1 showed the limitation of subword tokenization methods. Future work could explore methods which do not uses subword tokenization such as byT5 (Xue et al., 2022).

10 Conclusion

In this work, we showed that a translation model such as NLLB can perform the task of intent classification and slot filling. Because of tokenization issues, it is, however, suboptimal with non-spaced languages.

Moreover, we showed that artificially increasing the training sets' size leads to improved performance. Unfortunately, we also show that this added data can overlap with the existing test set, distorting the true evaluation of these models. The normal way to overcome this problem is to remove the overlap from the training set. However, deciding on what constitutes an overlap remains an open question. Therefore, we argued that the data overlap is not the problem — the evaluation metric is. As a result, we proposed two evaluation metrics that control the training/test overlap. Both metrics reveal that the model trained on overlapped data improves the results on non-overlapped data. However, our analysis also reveals that these models struggle to beat random chance when evaluated on utterances dissimilar to the training set.

Acknowledgement

We thank the reviewers for their helpful feedback. This research received funding from the Flemish Government under the *Onderzoeksprogramma Artificiële Intelligentie (AI) Vlaanderen* programme.

References

Amjad Almahairi, Nicolas Ballas, Tim Cooijmans, Yin Zheng, Hugo Larochelle, and Aaron Courville. 2016. Dynamic capacity networks. In *Proceedings of the 33rd International Conference on International Conference on Machine Learning - Volume 48*, ICML'16, page 2091–2100. JMLR.org.

Emanuele Bastianelli, Andrea Vanzo, Pawel Swietojanski, and Verena Rieser. 2020. SLURP: A spoken language understanding resource package. In *Proceedings of the 2020 Conference on Empirical Methods in Natural Language Processing (EMNLP)*, pages 7252–7262, Online. Association for Computational Linguistics.

Valentina Bellomaria, Giuseppe Castellucci, Andrea Favalli, and Raniero Romagnoli. 2019. Almawave-slu: A new dataset for slu in italian. *arXiv*, abs/1907.07526.

Yoshua Bengio, Nicholas Léonard, and Aaron C. Courville. 2013. Estimating or propagating gradients through stochastic neurons for conditional computation. *CoRR*, abs/1308.3432.

Tom Brown, Benjamin Mann, Nick Ryder, Melanie Subbiah, Jared D Kaplan, Prafulla Dhariwal, Arvind Neelakantan, Pranav Shyam, Girish Sastry, Amanda Askell, Sandhini Agarwal, Ariel Herbert-Voss, Gretchen Krueger, Tom Henighan, Rewon Child, Aditya Ramesh, Daniel Ziegler, Jeffrey Wu, Clemens Winter, Chris Hesse, Mark Chen, Eric Sigler, Mateusz Litwin, Scott Gray, Benjamin Chess, Jack Clark, Christopher Berner, Sam McCandlish, Alec Radford, Ilya Sutskever, and Dario Amodei. 2020. Language models are few-shot learners. In *Advances in Neural Information Processing Systems*, volume 33, pages 1877–1901. Curran Associates, Inc.

Giuseppe Castellucci, Valentina Bellomaria, Andrea Favalli, and Raniero Romagnoli. 2019. Multi-lingual intent detection and slot filling in a joint bert-based model. *arXiv*, abs/1907.02884.

Alexis Conneau, Kartikay Khandelwal, Naman Goyal, Vishrav Chaudhary, Guillaume Wenzek, Francisco Guzmán, Edouard Grave, Myle Ott, Luke Zettlemoyer, and Veselin Stoyanov. 2020. Unsupervised cross-lingual representation learning at scale. In *Proceedings of the 58th Annual Meeting of the Association for Computational Linguistics*, pages 8440–8451, Online. Association for Computational Linguistics.

Mai Hoang Dao, Thinh Hung Truong, and Dat Quoc Nguyen. 2021. Intent detection and slot filling for vietnamese. *arXiv*, abs/2104.02021.

Maxime De Bruyn, Ehsan Lotfi, Jeska Buhmann, and Walter Daelemans. 2021. MFAQ: a multilingual FAQ dataset. In *Proceedings of the 3rd Workshop on Machine Reading for Question Answering*, pages 1–13, Punta Cana, Dominican Republic. Association for Computational Linguistics.

Angela Fan, Shruti Bhosale, Holger Schwenk, Zhiyi Ma, Ahmed El-Kishky, Siddharth Goyal, Mandeep Baines, Onur Celebi, Guillaume Wenzek, Vishrav Chaudhary, Naman Goyal, Tom Birch, Vitaliy Liptchinsky, Sergey Edunov, Edouard Grave, Michael Auli, and Armand Joulin. 2021. Beyond english-centric multilingual machine translation. *J. Mach. Learn. Res.*, 22(1).

Jack FitzGerald, Christopher Hench, Charith Peris, Scott Mackie, Kay Rottmann, Ana Sanchez, Aaron Nash, Liam Urbach, Vishesh Kakarala, Richa Singh, Swetha Ranganath, Laurie Crist, Misha Britan, Wouter Leeuwis, Gokhan Tur, and Prem Natarajan. 2022. Massive: A 1m-example multilingual natural language understanding dataset with 51 typologically-diverse languages. *arXiv*, abs/2204.08582.

Yu Gong, Xusheng Luo, Yu Zhu, Wenwu Ou, Zhao Li, Muhua Zhu, Kenny Q. Zhu, Lu Duan, and Xi Chen. 2019. Deep cascade multi-task learning for slot filling in online shopping assistant. In *Proceedings of the Thirty-Third AAAI Conference on Artificial Intelligence and Thirty-First Innovative Applications of Artificial Intelligence Conference and Ninth AAAI Symposium on Educational Advances in Artificial Intelligence*, AAAI'19/IAAI'19/EAAI'19. AAAI Press.

Naman Goyal, Cynthia Gao, Vishrav Chaudhary, Peng-Jen Chen, Guillaume Wenzek, Da Ju, Sanjana Krishnan, Marc'Aurelio Ranzato, Francisco Guzmán, and Angela Fan. 2022. The Flores-101 evaluation benchmark for low-resource and multilingual machine translation. *Transactions of the Association for Computational Linguistics*, 10:522–538.

Sonal Gupta, Rushin Shah, Mrinal Mohit, Anuj Kumar, and Mike Lewis. 2018. Semantic parsing for task oriented dialog using hierarchical representations. In *Proceedings of the 2018 Conference on Empirical Methods in Natural Language Processing*, pages 2787–2792, Brussels, Belgium. Association for Computational Linguistics.

Xiaodong He, Li Deng, Dilek Hakkani-Tur, and Gokhan Tur. 2013. Multi-style adaptive training for robust cross-lingual spoken language understanding. In *2013 IEEE International Conference on Acoustics, Speech and Signal Processing*, pages 8342–8346.

Kevin Heffernan, Onur Çelebi, and Holger Schwenk. 2022. Bitext mining using distilled sentence representations for low-resource languages. *arXiv*, arxiv.2205.12654.

Geoffrey Hinton, Oriol Vinyals, and Jeff Dean. 2015. Distilling the knowledge in a neural network. *arXiv*, abs/1503.02531.

Taku Kudo and John Richardson. 2018. SentencePiece: A simple and language independent subword tokenizer and detokenizer for neural text processing. In *Proceedings of the 2018 Conference on Empirical Methods in Natural Language Processing: System Demonstrations*, pages 66–71, Brussels, Belgium. Association for Computational Linguistics.

Haoran Li, Abhinav Arora, Shuohui Chen, Anchit Gupta, Sonal Gupta, and Yashar Mehdad. 2021. MTOP: A comprehensive multilingual task-oriented semantic parsing benchmark. In *Proceedings of the 16th Conference of the European Chapter of the Association for Computational Linguistics: Main Volume*, pages 2950–2962, Online. Association for Computational Linguistics.

Bing Liu and Ian Lane. 2016. Attention-based recurrent neural network models for joint intent detection and slot filling. *Interspeech 2016*, pages 685–689.

Yinhan Liu, Jiatao Gu, Naman Goyal, Xian Li, Sergey Edunov, Marjan Ghazvininejad, Mike Lewis, and Luke Zettlemoyer. 2020. Multilingual denoising pre-training for neural machine translation. *arXiv*, abs/2001.08210.

Yinhan Liu, Myle Ott, Naman Goyal, Jingfei Du, Mandar Joshi, Danqi Chen, Omer Levy, Mike Lewis, Luke Zettlemoyer, and Veselin Stoyanov. 2019. Roberta: A robustly optimized bert pretraining approach. *arXiv*, abs/1907.11692.

Michael McTear. 2020. Conversational ai: dialogue systems, conversational agents, and chatbots. *Synthesis Lectures on Human Language Technologies*, 13(3):1–251.

Grégoire Mesnil, Xiaodong He, Li Deng, and Yoshua Bengio. 2013. Investigation of recurrent-neural-network architectures and learning methods for spoken language understanding. In *INTERSPEECH*.

Marie-Jean Meurs, Frédéric Duvert, Frédéric Béchet, Fabrice Lefèvre, and Renato de Mori. 2008. Semantic frame annotation on the French MEDIA corpus. In *Proceedings of the Sixth International Conference on Language Resources and Evaluation (LREC'08)*, Marrakech, Morocco. European Language Resources Association (ELRA).

Massimo Nicosia, Zhongdi Qu, and Yasemin Altun. 2021. Translate & Fill: Improving zero-shot multilingual semantic parsing with synthetic data. In *Findings of the Association for Computational Linguistics: EMNLP 2021*, pages 3272–3284, Punta Cana, Dominican Republic. Association for Computational Linguistics.

NLLB Team, Marta R. Costa-jussà, James Cross, Onur Çelebi, Maha Elbayad, Kenneth Heafield, Kevin Heffernan, Elahe Kalbassi, Janice Lam, Daniel Licht, Jean Maillard, Anna Sun, Skyler Wang, Guillaume Wenzek, Al Youngblood, Bapi Akula, Loic Barrault, Gabriel Mejia Gonzalez, Prangthip Hansanti, John Hoffman, Semarley Jarrett, Kaushik Ram Sadagopan, Dirk Rowe, Shannon Spruit, Chau Tran, Pierre Andrews, Necip Fazil Ayan, Shruti Bhosale, Sergey Edunov, Angela Fan, Cynthia Gao, Vedanuj Goswami, Francisco Guzmán, Philipp Koehn, Alexandre Mourachko, Christophe Ropers, Safiyyah Saleem, Holger Schwenk, and Jeff Wang. 2022. No language left behind: Scaling human-centered machine translation. *arXiv*, abs/2207.04672.

Adam Paszke, Sam Gross, Francisco Massa, Adam Lerer, James Bradbury, Gregory Chanan, Trevor Killeen, Zeming Lin, Natalia Gimelshein, Luca

Antiga, Alban Desmaison, Andreas Kopf, Edward Yang, Zachary DeVito, Martin Raison, Alykhan Tejani, Sasank Chilamkurthy, Benoit Steiner, Lu Fang, Junjie Bai, and Soumith Chintala. 2019. Pytorch: An imperative style, high-performance deep learning library. In H. Wallach, H. Larochelle, A. Beygelzimer, F. d'Alché-Buc, E. Fox, and R. Garnett, editors, *Advances in Neural Information Processing Systems 32*, pages 8024–8035. Curran Associates, Inc.

P. J. Price. 1990. Evaluation of spoken language systems: the ATIS domain. In *Speech and Natural Language: Proceedings of a Workshop Held at Hidden Valley, Pennsylvania, June 24-27,1990.*

Colin Raffel, Noam Shazeer, Adam Roberts, Katherine Lee, Sharan Narang, Michael Matena, Yanqi Zhou, Wei Li, and Peter J. Liu. 2020. Exploring the limits of transfer learning with a unified text-to-text transformer. *Journal of Machine Learning Research*, 21(140):1–67.

Samyam Rajbhandari, Jeff Rasley, Olatunji Ruwase, and Yuxiong He. 2020. Zero: Memory optimizations toward training trillion parameter models. In *Proceedings of the International Conference for High Performance Computing, Networking, Storage and Analysis*, SC '20. IEEE Press.

Evgeniia Razumovskaia, Goran Glavas, Olga Majewska, Edoardo M Ponti, Anna Korhonen, and Ivan Vulic. 2022. Crossing the conversational chasm: A primer on natural language processing for multilingual task-oriented dialogue systems. *Journal of Artificial Intelligence Research*, 74:1351–1402.

Nils Reimers and Iryna Gurevych. 2020. Making monolingual sentence embeddings multilingual using knowledge distillation. In *Proceedings of the 2020 Conference on Empirical Methods in Natural Language Processing (EMNLP)*, pages 4512–4525, Online. Association for Computational Linguistics.

Sebastian Schuster, Sonal Gupta, Rushin Shah, and Mike Lewis. 2019. Cross-lingual transfer learning for multilingual task oriented dialog. In *Proceedings of the 2019 Conference of the North American Chapter of the Association for Computational Linguistics: Human Language Technologies, Volume 1 (Long and Short Papers)*, pages 3795–3805, Minneapolis, Minnesota. Association for Computational Linguistics.

Skipper Seabold and Josef Perktold. 2010. statsmodels: Econometric and statistical modeling with python. In *9th Python in Science Conference.*

Raymond Hendy Susanto and Wei Lu. 2017. Neural architectures for multilingual semantic parsing. In *Proceedings of the 55th Annual Meeting of the Association for Computational Linguistics (Volume 2: Short Papers)*, pages 38–44, Vancouver, Canada. Association for Computational Linguistics.

Shyam Upadhyay, Manaal Faruqui, Gokhan Tür, Hakkani-Tür Dilek, and Larry Heck. 2018. (almost) zero-shot cross-lingual spoken language understanding. In *2018 IEEE International Conference on Acoustics, Speech and Signal Processing (ICASSP)*, pages 6034–6038.

Ashish Vaswani, Noam Shazeer, Niki Parmar, Jakob Uszkoreit, Llion Jones, Aidan N Gomez, Ł ukasz Kaiser, and Illia Polosukhin. 2017. Attention is all you need. In *Advances in Neural Information Processing Systems*, volume 30. Curran Associates, Inc.

Thomas Wolf, Lysandre Debut, Victor Sanh, Julien Chaumond, Clement Delangue, Anthony Moi, Pierric Cistac, Tim Rault, Remi Louf, Morgan Funtowicz, Joe Davison, Sam Shleifer, Patrick von Platen, Clara Ma, Yacine Jernite, Julien Plu, Canwen Xu, Teven Le Scao, Sylvain Gugger, Mariama Drame, Quentin Lhoest, and Alexander Rush. 2020. Transformers: State-of-the-art natural language processing. In *Proceedings of the 2020 Conference on Empirical Methods in Natural Language Processing: System Demonstrations*, pages 38–45, Online. Association for Computational Linguistics.

Linting Xue, Aditya Barua, Noah Constant, Rami Al-Rfou, Sharan Narang, Mihir Kale, Adam Roberts, and Colin Raffel. 2022. ByT5: Towards a token-free future with pre-trained byte-to-byte models. *Transactions of the Association for Computational Linguistics*, 10:291–306.

Linting Xue, Noah Constant, Adam Roberts, Mihir Kale, Rami Al-Rfou, Aditya Siddhant, Aditya Barua, and Colin Raffel. 2021. mT5: A massively multilingual pre-trained text-to-text transformer. In *Proceedings of the 2021 Conference of the North American Chapter of the Association for Computational Linguistics: Human Language Technologies*, pages 483–498, Online. Association for Computational Linguistics.

Huiyun Yang, Huadong Chen, Hao Zhou, and Lei Li. 2022. Enhancing cross-lingual transfer by manifold mixup. *arXiv*, abs/2205.04182.

Yinfei Yang, Daniel Cer, Amin Ahmad, Mandy Guo, Jax Law, Noah Constant, Gustavo Hernandez Abrego, Steve Yuan, Chris Tar, Yun-hsuan Sung, Brian Strope, and Ray Kurzweil. 2020. Multilingual universal sentence encoder for semantic retrieval. In *Proceedings of the 58th Annual Meeting of the Association for Computational Linguistics: System Demonstrations*, pages 87–94, Online. Association for Computational Linguistics.

Qingyu Zhang, Xiaoyu Shen, Ernie Chang, Jidong Ge, and Pengke Chen. 2022. Mdia: A benchmark for multilingual dialogue generation in 46 languages. *arXiv*, abs/2208.13078.

Weinan Zhang, Zhigang Chen, Wanxiang Che, Guoping Hu, and Ting Liu. 2017. The first evaluation of chinese human-computer dialogue technology. *CoRR*, abs/1709.10217.

Bo Zheng, Li Dong, Shaohan Huang, Wenhui Wang, Zewen Chi, Saksham Singhal, Wanxiang Che, Ting Liu, Xia Song, and Furu Wei. 2021. Consistency regularization for cross-lingual fine-tuning. In *Proceedings of the 59th Annual Meeting of the Association for Computational Linguistics and the 11th International Joint Conference on Natural Language Processing (Volume 1: Long Papers)*, pages 3403–3417, Online. Association for Computational Linguistics.

Su Zhu and Kai Yu. 2017. Encoder-decoder with focus-mechanism for sequence labelling based spoken language understanding. In *2017 IEEE International Conference on Acoustics, Speech and Signal Processing (ICASSP)*, pages 5675–5679. IEEE.

A Distribution of Intents & Slots

We list in Table 6 the distribution of intents across the three datasets. Table 7 shows the distribution of slots across the three datasets.

B Logistic Regression by Languages

We list the results of the logistic regression by language in Table 8.

Intent	MASSIVE	Generated	Synthetic
calendar_set	7,0%	2,6%	3,6%
play_music	5,5%	2,2%	3,3%
weather_query	5,0%	2,0%	3,0%
calendar_query	4,9%	2,2%	2,4%
general_quirky	4,8%	1,7%	5,2%
qa_factoid	4,7%	2,0%	8,3%
news_query	4,4%	2,1%	2,5%
email_query	3,6%	2,2%	12,0%
email_sendemail	3,1%	2,4%	11,1%
datetime_query	3,0%	1,5%	1,4%
calendar_remove	2,7%	2,1%	1,1%
play_radio	2,5%	2,1%	1,5%
social_post	2,5%	2,4%	8,7%
qa_definition	2,3%	2,2%	3,7%
transport_query	2,0%	2,3%	1,1%
cooking_recipe	1,8%	2,2%	1,2%
lists_query	1,7%	1,5%	1,0%
play_podcasts	1,7%	1,5%	1,0%
recommendation_events	1,7%	2,0%	0,9%
alarm_set	1,6%	1,8%	0,6%
lists_createoradd	1,5%	1,7%	0,6%
recommendation_locations	1,5%	2,3%	0,9%
lists_remove	1,4%	1,7%	0,9%
music_query	1,3%	1,3%	0,6%
iot_hue_lightoff	1,3%	1,3%	0,6%
qa_stock	1,3%	2,5%	2,7%
play_audiobook	1,3%	2,0%	0,3%
qa_currency	1,2%	2,2%	3,3%
takeaway_order	1,2%	2,1%	0,4%
alarm_query	1,1%	1,3%	0,2%
email_querycontact	1,1%	2,0%	3,3%
transport_ticket	1,1%	1,8%	0,6%
iot_hue_lightchange	1,1%	2,1%	0,7%
iot_coffee	1,1%	1,2%	0,5%
takeaway_query	1,1%	1,8%	0,5%
transport_traffic	1,0%	1,8%	0,4%
music_likeness	1,0%	1,5%	0,5%
play_game	1,0%	1,7%	0,7%
audio_volume_up	1,0%	1,2%	0,1%
audio_volume_mute	1,0%	1,5%	0,3%
social_query	0,9%	2,0%	2,8%
transport_taxi	0,9%	1,9%	0,5%
iot_cleaning	0,8%	1,4%	0,4%
alarm_remove	0,7%	1,8%	0,2%
qa_maths	0,7%	1,7%	0,8%
iot_hue_lightup	0,7%	1,3%	0,4%
iot_hue_lightdim	0,7%	1,4%	0,4%
general_joke	0,6%	1,3%	0,3%
recommendation_movies	0,6%	2,0%	0,4%
email_addcontact	0,5%	1,3%	1,4%
iot_wemo_off	0,5%	0,8%	0,2%
datetime_convert	0,5%	1,6%	0,2%
audio_volume_down	0,5%	1,1%	0,1%
music_settings	0,4%	0,9%	0,2%
iot_wemo_on	0,4%	1,0%	0,2%
general_greet	0,2%	0,2%	
iot_hue_lighton	0,2%	1,0%	0,1%
audio_volume_other	0,2%	0,6%	0,0%
music_dislikeness	0,1%	0,9%	0,1%
cooking_query	0,0%	0,0%	0,0%

Table 6: Distribution of intents across the three datasets. Generated represents the utterances generated by GPT-3, while synthetic represents the synthetic training set of SLURP.

Intent	MASSIVE	Generated	Synthetic
date	16,0%	10,8%	10,7%
place_name	9,6%	10,6%	8,0%
event_name	8,8%	4,3%	5,5%
person	7,6%	5,4%	17,2%
time	7,0%	5,8%	4,1%
media_type	4,2%	5,4%	9,5%
business_name	3,4%	5,7%	7,6%
weather_descriptor	2,8%	1,1%	1,5%
transport_type	2,8%	5,0%	1,2%
food_type	2,6%	4,2%	1,4%
relation	2,2%	2,3%	4,8%
timeofday	2,1%	2,0%	1,3%
artist_name	2,0%	0,8%	1,2%
device_type	2,0%	3,4%	1,1%
definition_word	2,0%	2,0%	3,5%
currency_name	1,9%	3,8%	5,7%
house_place	1,7%	3,8%	0,8%
list_name	1,7%	1,8%	0,9%
business_type	1,7%	2,8%	0,8%
news_topic	1,6%	0,7%	1,1%
music_genre	1,6%	0,9%	1,0%
player_setting	1,4%	2,1%	0,5%
radio_name	1,2%	1,1%	0,9%
song_name	1,1%	0,3%	0,7%
order_type	0,9%	1,6%	0,3%
color_type	0,9%	1,7%	0,4%
game_name	0,8%	1,3%	0,6%
general_frequency	0,7%	0,3%	0,4%
personal_info	0,7%	1,2%	2,0%
audiobook_name	0,6%	0,9%	0,2%
podcast_descriptor	0,6%	0,6%	0,3%
meal_type	0,6%	0,4%	0,4%
playlist_name	0,5%	0,1%	0,3%
podcast_name	0,5%	0,4%	0,3%
time_zone	0,5%	1,1%	0,2%
app_name	0,4%	0,3%	0,1%
change_amount	0,4%	0,9%	0,1%
music_descriptor	0,4%	0,2%	0,2%
joke_type	0,3%	0,8%	0,2%
email_folder	0,3%	0,2%	0,9%
email_address	0,3%	0,4%	1,4%
transport_agency	0,3%	0,5%	0,2%
coffee_type	0,2%	0,2%	0,1%
ingredient	0,2%	0,1%	0,1%
cooking_type	0,1%	0,1%	0,1%
movie_name	0,1%	0,1%	0,1%
movie_type	0,1%	0,2%	0,0%
transport_name	0,1%	0,1%	0,1%
drink_type	0,1%	0,1%	0,0%
alarm_type	0,1%	0,1%	0,0%
transport_descriptor	0,1%	0,0%	0,0%
audiobook_author	0,1%	0,2%	0,0%
sport_type	0,0%	0,0%	0,0%
music_album	0,0%		0,0%
game_type	0,0%	0,0%	0,0%

Table 7: Distribution of slots across the three datasets. Generated represents the utterances generated by GPT-3, while synthetic represents the synthetic training set of SLURP.

language	β_0	β_1	R_2	$f(x=0)$	$f(x=0.5)$	$f(x=1)$
all	-0.69	3.14	0.08	0.33	0.71	0.92
af-ZA	-0.98	4.01	0.11	0.27	0.74	0.95
am-ET	-0.46	3.09	0.06	0.39	0.75	0.93
ar-SA	-0.58	3.01	0.07	0.36	0.72	0.92
az-AZ	-0.55	3.24	0.08	0.37	0.75	0.94
bn-BD	-1.27	3.71	0.10	0.22	0.64	0.92
cy-GB	-0.66	3.37	0.08	0.34	0.74	0.94
da-DK	-0.95	4.13	0.12	0.28	0.75	0.96
de-DE	-0.65	3.58	0.09	0.34	0.76	0.95
el-GR	-0.92	3.64	0.09	0.28	0.71	0.94
en-US	-1.45	4.93	0.21	0.19	0.73	0.97
es-ES	-0.60	2.99	0.07	0.36	0.71	0.92
fa-IR	-0.96	2.70	0.06	0.28	0.60	0.85
fi-FI	-0.86	3.80	0.10	0.30	0.74	0.95
fr-FR	-0.37	2.65	0.05	0.41	0.72	0.91
he-IL	-0.72	3.44	0.08	0.33	0.73	0.94
hi-IN	-0.76	3.10	0.08	0.32	0.69	0.91
hu-HU	-0.55	3.25	0.08	0.37	0.75	0.94
hy-AM	-1.05	3.35	0.08	0.26	0.65	0.91
id-ID	-0.67	3.33	0.08	0.34	0.73	0.93
is-IS	-0.56	3.19	0.07	0.36	0.74	0.93
it-IT	-0.46	2.82	0.06	0.39	0.72	0.91
ja-JP	-0.48	2.77	0.06	0.38	0.71	0.91
jv-ID	-0.34	2.95	0.06	0.42	0.76	0.93
ka-GE	-0.46	2.59	0.06	0.39	0.70	0.89
km-KH	-0.23	1.62	0.03	0.44	0.64	0.80
kn-IN	-0.94	2.55	0.05	0.28	0.58	0.83
ko-KR	-0.49	3.42	0.08	0.38	0.77	0.95
lv-LV	-0.81	3.62	0.09	0.31	0.73	0.94
ml-IN	-1.39	3.64	0.10	0.20	0.61	0.90
mn-MN	-0.79	3.32	0.07	0.31	0.70	0.93
ms-MY	-0.77	3.55	0.08	0.32	0.73	0.94
my-MM	-0.97	4.12	0.08	0.27	0.75	0.96
nb-NO	-0.72	3.65	0.09	0.33	0.75	0.95
nl-NL	-0.80	3.71	0.10	0.31	0.74	0.95
pl-PL	-0.52	2.65	0.06	0.37	0.69	0.89
pt-PT	-0.56	3.05	0.07	0.36	0.72	0.92
ro-RO	-0.36	3.00	0.06	0.41	0.76	0.93
ru-RU	-0.47	3.12	0.07	0.38	0.75	0.93
sl-SL	-0.63	3.25	0.08	0.35	0.73	0.93
sq-AL	-0.54	3.04	0.07	0.37	0.73	0.92
sv-SE	-0.51	3.53	0.09	0.37	0.78	0.95
sw-KE	-0.89	3.26	0.08	0.29	0.68	0.91
ta-IN	-0.70	3.20	0.07	0.33	0.71	0.92
te-IN	-0.65	2.18	0.04	0.34	0.61	0.82
th-TH	-0.66	2.61	0.06	0.34	0.66	0.88
tl-PH	-1.12	3.72	0.09	0.25	0.68	0.93
tr-TR	-0.71	3.53	0.09	0.33	0.74	0.94
ur-PK	-0.80	3.30	0.08	0.31	0.70	0.92
vi-VN	-1.72	3.78	0.10	0.15	0.54	0.89
zh-CN	-0.42	2.35	0.06	0.40	0.68	0.87
zh-TW	-0.56	1.97	0.05	0.36	0.61	0.80

Table 8: Logistic regression results by language

The Massively Multilingual Natural Language Understanding 2022 (MMNLU-22) Workshop and Competition

Chris Hench
henchc@amazon.com

Charith Peris
perisc@amazon.com

Jack FitzGerald
jgmf@amazon.com

Kay Rottmann
krrottm@amazon.com

Abstract

Despite recent progress in Natural Language Understanding (NLU), the creation of multilingual NLU systems remains a challenge. It is common to have NLU systems limited to a subset of languages due to lack of available data. They also often vary widely in performance. We launch a three-phase approach to address the limitations in NLU and help propel NLU technology to new heights. We release a 52 language dataset called the Multilingual Amazon SLU resource package (SLURP) for Slot-filling, Intent classification, and Virtual assistant Evaluation, or MASSIVE, in an effort to address parallel data availability for voice assistants. We organize the Massively Multilingual NLU 2022 Challenge to provide a competitive environment and push the state-of-the art in the transferability of models into other languages. Finally, we host the first Massively Multilingual NLU workshop which brings these components together. The MMNLU workshop seeks to advance the science behind multilingual NLU by providing a platform for the presentation of new research in the field and connecting teams working on this research direction. This paper summarizes the dataset, workshop and the competition and the findings of each phase.

1 Introduction

According to a 2020 study by Juniper Research (Juniper, 2020) it is expected that by 2024 there will be over 8 billion virtual assistants worldwide, the majority of which will be on smartphones. Additionally, over 100 million smart speakers have been sold, and virtual assistants continue to be integrated into new products. These devices have in common that humans interact with them via natural language interfaces. This development has significantly boosted research to advance natural language understanding. However, most natural language understanding work focuses on only a few of the more than 4,000 written languages in the world. The limitation is driven by the lack of labeled data, the expense associated with human-based quality assurance, model maintenance, update costs, and more. To overcome these hurdles, further research in the field of multilingual natural language understanding is needed to enable natural language understanding for currently not- or under-served languages. With NLLB Team et al. (2022) and related work, we have seen progress in recent years on the expansion of machine translation into the domain of under-served languages both by advancing science as well as creation of corpora in machine translation. However, in areas as NLU modeling for virtual assistants, many of these limitations still remain. The vision of this workshop is to address the limitations in NLU and help propel NLU technology into the 50-language, 100-language, and even the 1,000-language regime, both for production systems and for research endeavors, succinctly captured by our slogan, *Let's scale natural language understanding technology to every language on Earth!*. We do this via a three-pronged approach. First, we created and released the Multilingual Amazon SLU resource package (SLURP) for Slot-filling, Intent classification, and Virtual assistant Evaluation, or MASSIVE dataset (FitzGerald et al., 2022), containing 1 million realistic, parallel, labeled virtual assistant text utterances spanning 51 languages. Second, we hosted the Massively Multilingual NLU (MMNLU) 2022 Challenge, a competition designed to advance massively multilingual NLU modeling. Finally, we organized the first MMNLU workshop to bring together researchers working in the field of NLU. By providing much needed labelled data, motivating multilingual NLU exploration and bringing NLU researchers together to share findings and spark further collaboration, we hope to push the state-of-the art in multilingual natural language understanding technology.

Proceedings of the Massively Multilingual Natural Language Understanding Workshop (MMNLU-22), pages 83 - 87
December 7, 2022 ©2022 Association for Computational Linguistics

2 Workshop overview

The first MMNLU 2022 workshop is co-located with EMNLP 2022 in Abu Dhabi. In our call for papers we asked for submissions relevant to the advancement of the field of multilingual NLP. We were particularly interested in submissions related to the shared tasks part of this workshops competition on the recently published MASSIVE (FitzGerald et al., 2022) dataset (see 4.1) or other multilingual data-sets. We sought work exploring multilingual representations, augmentation and preprocessing techniques, and more efficient models.

3 Paper Submissions

In total we received 12 submissions for the venue, of which 8 were accepted for presentation at the workshop. The papers being part of the proceedings for this workshop showed a wide variety of approaches to deal with the problem of massive multilingual systems. The investigations ranged from the evaluation and investigation of tokenization across languages, towards large language model or translation model based data augmentation as well as direct use of translation systems as a natural language understanding solution. Also, investigations how to minimize degradation on other languages when trained only on a small set of languages, as well as how to use consistency regularization as a way to boost performance were part of the submissions to this workshop. Another topic investigated by more than one submission was the investigation of code mixing and other cross lingual effects on natural language understanding performance. Furthermore we also received a paper describing an investigation of how to design templates when Seq2Seq generation is used as a solution for zero-shot cross-lingual tagging. Overall we are very grateful for the diverse set of research directions proposed in the submissions to this workshop.

4 The Massive Multilingual NLU 2022 Challenge

4.1 MASSIVE dataset

The MASSIVE dataset was localized across 50 languages from the original English data released in the SLURP NLU dataset (Bastianelli et al., 2020). Unique ids were preserved to yield a parallel corpus, allowing for various natural language understanding tasks beyond intent and slot recognition, such as machine translation. The dataset comprises 60 intents and 55 slot types across 18 domains. The released dataset was partitioned into 587k training utterances, 104k development utterances, and 152k test utterances, also preserving the split used by the original SLURP dataset. For the MMNLU-22 competition, an additional 153k utterances were held-out for the leaderboard. The held-out utterances were created by professionals manually paraphrasing a random sample of SLURP utterances in English, which were subsequently localized along with the original dataset. This resulted in more challenging utterances for NLU, with 49% more slots per utterance on average.

4.2 The Competition

The MMNLU 2022 Challenge is designed to advance the state of the art of massively multilingual NLU, in which a single model can understand and parse text inputs from many different languages. The competition is based on the recently published MASSIVE (FitzGerald et al., 2022) dataset (see Section 4.1).

The competition consisted of two tasks; namely (1) the Full Dataset Task (4.2.1) and (2) the Zero Shot task (4.2.2). The competition was geared towards two awards: the top-scoring award for the system with the best performance, which was awarded separately for each task, and the organizer's choice award, which was awarded after considering overall submissions. The competition ran from July 24, 2022 until September 3, 2022.

4.2.1 Full Dataset Task

For the Full Dataset Task, a single model, trained on all languages of MASSIVE and according to the given split and hold out data constraints, was evaluated on all languages of the MASSIVE hidden evaluation set consisting of 3,000 utterances per locale. All encoder-only models were required to have fewer than 350M parameters and all sequence-to-sequence models fewer than 700M parameters, including word embeddings. We permitted any data to be used for training, but they must have been publicly available. If not publicly available, or a third party service such as machine translation was utilized, we requested these data be made public. Only the training split was permitted for training, development and test partitions were not. No use of the text in the evaluation set for training was permitted.

4.2.2 Zero Shot Task

For the Zero Shot Task, a single model, trained on only the English training partition of MASSIVE and according to all other given constraints in the Full Dataset Task, was evaluated on every language except English in the MASSIVE hidden evaluation set.

4.3 Top-scoring award

The top-scoring award was intended to encourage teams to create models within the constraints outlined for the competition (see Sections 4.2.1 and 4.2.2) while demonstrating the best performance for a given task. The performance of submissions to the top-scoring award were evaluated based on Exact Match Accuracy (EMA) of the intent and slot labels in the predicted utterances provided by the competitor, when matched against the labels in the ground-truth (i.e., the MASSIVE hidden evaluation set).

The two teams that achieved first place on the leaderboard for the two tasks were announced as winners (see Sections 4.6.1 and 4.6.2 for winning teams and descriptions of their approaches).

4.4 Organizer's choice award

In addition to simple utility, we also wanted to encourage creative approaches to solving the problem of intent classification and named entity recognition. The organizers' choice award was based primarily on the assessment of the promise of a given approach, and not purely on its leaderboard performance. The assessment was made by a panel of reviewers that consisted of the Program Committee and Organizer's of MMNLU. See Section 4.6.3 for the winning team and a description of their approach.

4.5 Leaderboard

Our leaderboard[1] was setup on eval.ai (Yadav et al., 2019) a centralized platform that hosts Artificial Intelligence (AI) challenges across the globe with the intention of supporting better benchmarking in AI. eval.ai allows for two types of challenge to be hosted on their servers, namely (1) prediction upload challenges, and (2) code upload challenges.

The MMNLU competition took the form of a prediction upload challenge where competitors were required to train their own models, score them on a hidden evaluation set and then upload the predictions onto the eval.ai challenge. Once uploaded the predictions were run through an evaluation script. The evaluation script was designed to calculate exact match accuracy (EMA), intent accuracy, slot F1, the EMA of the highest performing language and the EMA of the lowest performing language, against the ground truth of the hidden evaluation set.

4.6 Submissions

We received seven submissions to the Full Dataset task and eight submissions to the ZeroShot task. A total of 11 research teams participated across the two tasks. A top-scoring award was given to the two teams that led the leaderboard for each task, and the organizer's choice award was given based on the assessment of the promise of a given approach. We briefly describe the winning submissions in the following sections.

4.6.1 Team HIT-SCIR

Team `HIT-SCIR` won the top-scoring award for the Full Dataset Task (see Sections 4.3 and 4.2.1). They used a consistency regularization approach with a hybrid data augmentation strategy that included machine translation and subword sampling[2]. Consistency regularization, applied via symmetric Kullback-Leibler divergence, was used to encourage the predicted distributions for an example and its semantically equivalent augmentation to agree with each other.

For the intent detection task, the original example and the predicted distributions of the augmentation data from both strategies are directly aligned. For the slot filling task, the original example can only be aligned with the predicted distribution from the subword sampling augmentation strategy. Slot consistency is ignored when using augmented data from machine translation.

For the Full Dataset task, they used the mT5-base text-to-text model presented by FitzGerald et al. (2022), which contains 580M parameters, including 190M embedding parameters. They used examples with the same id in different languages as machine translation augmentation (see FitzGerald et al. 2022 for details on the MASSIVE dataset id).

For the Zero Shot task, they used the XLM-align-base model, which contains 270M parameters, in-

[1] https://eval.ai/web/challenges/challenge-page/1697/leaderboard

[2] Subword sampling is to apply the on-the-fly subword sampling algorithm in the unigram language model to generate multiple tokenized subword sequences

cluding 190M embedding parameters, and a simple two-layer classification head for both intent detection and slot-filling tasks. They used commercial translation APIs to obtain slot-aligned translations and intent-aligned translations.

The consistency regularization-based method does not introduce any additional parameters into the model.

For a detailed description of their submission, refer Team `HIT-SCIR`'s work (Zheng et al., 2022).

4.6.2 Team FabT5

Team `FabT5` won the top-scoring award for the Zero Shot Task (see Sections 4.3 and 4.2.2). They used ByT5 base (Xue et al., 2022), a text-to-text model that takes an input query and outputs a full interpretation composed of the intent, relevant slot labels, and their corresponding slot values. ByT5 base has the same number of parameters as mT5 base (582M), but distributed differently. While mT5 base has 66% of its parameters allocated for the embeddings, ByT5 base has only 0.1% of its parameters allocated for the same purpose. They trained the ByT5 model to predict an MTOP-style interpretation from a given query. These predictions were then converted back into the intent and annotated utterance field formats in the MASSIVE dataset (FitzGerald et al., 2022).

In addition, they found that prepending both query and target interpretation with the language string of the query was slightly helpful.

For the zero-shot submission the team trained the model using the English MASSIVE training split. To obtain data in the target languages, they translated the English queries using the Google Translate API and projected the slot annotations from the original English queries to the corresponding translations. To project the annotations, they used the Translate-and-Fill approach, where an mT5 filler model trained on the English queries was used to project the labels to the translations in a zero-shot fashion. The English train partition was the only one used for zero-shot training. The English validation set was used to select the best checkpoint for inference. No hyperparameter tuning was performed. A fixed learning rate of 0.0001 and a batch size of 128 was used for training.

For a detailed description of their submission, refer Team `FabT5`'s work (Nicosia and Piccinno, 2022).

4.6.3 Team bolleke

Team `bolleke` won the organizer's choice award (Section 4.4). They repurposed a translation model for the task of intent detection and slot filling. The existing dataset was first expanded by generating new training examples via the use of two fine-tuned GPT-3 models. One variant of GPT-3 (13B) was used to generate utterances conditioned on the intent, and a second variant was used to do the intent and slot filling task. Using the two models added confidence via intent agreement. The team generated 20K (English) examples using this method. Next, a translation model was trained on the MASSIVE training partition to translate English annotated utterances into the 50 available languages, resulting in an augmented dataset of an additional 1M examples. For both pre-training and fine-tuning, the team used an NLLB-200's (No Language Left Behind) distilled 600M parameter variant (NLLB Team et al., 2022). The objective of the pre-training step was a cross-lingual translation task, with intent detection and slot filling (for example, they trained the model to translate *"wat is het weer in new york"* to *"weather_query|quel est le temps à [place_name : new york]"*). The fine-tuning was done on the augmented dataset for another 50K steps with the objective of predicting the annotated utterance in the same language (for example, they train the model to output *"weather_query|what is the weather in [place_name : new york]"* given the input *"what is the weather in new york"*).

Training was done on DeepSpeed with a batch size of 56 and a max length of 64. The input is always the raw utterance and the output the concatenation of the intent and the annotated utterance. Learning rates of 0.0001 and 0.00005 were used for the pre-training and fine-tuning respectively. Given the size of the dataset and computing resources involved they did not engage in hyper-parameter tuning.

For a detailed description of their submission, refer Team `bolleke`'s work (Bruyn et al., 2022).

5 Conclusions

This paper presents an overview of the first MMNLU workshop collocated with EMNLP 2022. The paper submissions and competition entries showed encouraging progress in the field of multilingual NLU. We hope that the findings presented as well as the collaborations initiated at this workshop, drives more progress in the field.

Acknowledgements

We would like to thank the program committee and the reviewers for their important contributions to the organization of the first MMNLU workshop.

References

Emanuele Bastianelli, Andrea Vanzo, Pawel Swietojanski, and Verena Rieser. 2020. Slurp: A spoken language understanding resource package.

Maxime De Bruyn, Ehsan Lotfi, Jeska Buhmann, and Walter Daelemans. 2022. Machine translation for multilingual intent detection and slots filling. In *EMNLP 2022 Massively Multilingual Natural Language Understanding (MMNLU)*.

Jack FitzGerald, Christopher Hench, Charith Peris, Scott Mackie, Kay Rottmann, Ana Sanchez, Aaron Nash, Liam Urbach, Vishesh Kakarala, Richa Singh, Swetha Ranganath, Laurie Crist, Misha Britan, Wouter Leeuwis, Gokhan Tur, and Prem Natarajan. 2022. Massive: A 1m-example multilingual natural language understanding dataset with 51 typologically-diverse languages.

Research Juniper. 2020. Number of voice assistant devices in use to overtake world population by 2024, reaching 8.4bn, led by smartphones.

Massimo Nicosia and Francesco Piccinno. 2022. Evaluating byte and wordpiece level models for massively multilingual semantic parsing. In *EMNLP 2022 Massively Multilingual Natural Language Understanding (MMNLU)*.

NLLB Team, Marta R. Costa-jussà, James Cross, Onur Çelebi, Maha Elbayad, Kenneth Heafield, Kevin Heffernan, Elahe Kalbassi, Janice Lam, Daniel Licht, Jean Maillard, Anna Sun, Skyler Wang, Guillaume Wenzek, Al Youngblood, Bapi Akula, Loic Barrault, Gabriel Mejia Gonzalez, Prangthip Hansanti, John Hoffman, Semarley Jarrett, Kaushik Ram Sadagopan, Dirk Rowe, Shannon Spruit, Chau Tran, Pierre Andrews, Necip Fazil Ayan, Shruti Bhosale, Sergey Edunov, Angela Fan, Cynthia Gao, Vedanuj Goswami, Francisco Guzmán, Philipp Koehn, Alexandre Mourachko, Christophe Ropers, Safiyyah Saleem, Holger Schwenk, and Jeff Wang. 2022. No language left behind: Scaling human-centered machine translation.

Linting Xue, Aditya Barua, Noah Constant, Rami Al-Rfou, Sharan Narang, Mihir Kale, Adam Roberts, and Colin Raffel. 2022. Byt5: Towards a token-free future with pre-trained byte-to-byte models. *Transactions of the Association for Computational Linguistics*, 10:291–306.

Deshraj Yadav, Rishabh Jain, Harsh Agrawal, Prithvijit Chattopadhyay, Taranjeet Singh, Akash Jain, Shivkaran Singh, Stefan Lee, and Dhruv Batra. 2019. Evalai: Towards better evaluation systems for ai agents. *ArXiv*, abs/1902.03570.

Bo Zheng, Zhouyang Li, Fuxuan Wei, Qiguang Chen, Libo Qin, and Wanxiang Che. 2022. Hit-scir at mmnlu-22: Consistency regularization for multilingual spoken language understanding. In *EMNLP 2022 Massively Multilingual Natural Language Understanding (MMNLU)*.

Author Index

Association for Computational Linguistics
209 N. Eighth Street
Stroudsburg, Pennsylvania 18360

ISBN 978-1-7138-9117-8

CURRAN ASSOCIATES INC.
proceedings.com